CAMBRIDGE MONOGRAPHS ON MECHANICS AND APPLIED MATHEMATICS

GENERAL EDITORS

G. K. BATCHELOR, PH.D.
Lecturer in Mathematics in the University of Cambridge

H. BONDI, M.A.
Lecturer in Mathematics in the University of Cambridge

GEOMETRICAL MECHANICS

AND

DE BROGLIE WAVES

GEOMETRICAL MECHANICS
AND
DE BROGLIE WAVES

BY

J. L. SYNGE, Sc.D., F.R.S.

Senior Professor, School of Theoretical Physics
Dublin Institute for Advanced Studies

CAMBRIDGE
AT THE UNIVERSITY PRESS
1954

CAMBRIDGE UNIVERSITY PRESS
Cambridge, New York, Melbourne, Madrid, Cape Town, Singapore,
São Paulo, Delhi, Dubai, Tokyo, Mexico City

Cambridge University Press
The Edinburgh Building, Cambridge CB2 8RU, UK

Published in the United States of America by Cambridge University Press, New York

www.cambridge.org
Information on this title: www.cambridge.org/9780521156882

First published 1954
First paperback edition 2010

A catalogue record for this publication is available from the British Library

ISBN 978-0-521-15688-2 Paperback

CONTENTS

IV Primitive Quantization

V Some Generalizations

PREFACE

Mathematical physicists are thoroughly familiar with Hamilton's dynamical method, but his optical method seems to be almost unknown except by those interested in the theory of optical instruments and electron microscopes, particularly the theory of aberrations. Certainly, although I had been familiar with the optical method for over twenty years, I had hardly thought of it outside that technical setting until quite recently I put this question to myself: 'Hamilton based his theory of rays and waves on a variational principle (Fermat's) in *space*, all three coordinates being treated on an equal footing; what sort of a theory would one get by applying Hamilton's method to such a variational principle in *space-time*, treating the four coordinates on an equal footing?'

To my surprise there emerged a general and completely relativistic theory of de Broglie waves—relativistic geometrical mechanics—with an attractive simplicity of form, lending itself readily to visualization in space-time, the rays appearing as curves in space-time (the histories of particles) and the waves as 3-spaces. It is the primary purpose of this book to develop this theory systematically, supplementing the theory which flows naturally from Hamilton's method with a process of 'primitive quantization' in order to be able to deal with the interference of material waves. I have tried to dilute the abstractness of the general theory by working out some particular examples in detail. In the Introduction I have set out what I conceive to be the status in physics of the subject as here developed.

I wish to thank Professor E. Schrödinger, and also the Scholars in the School of Theoretical Physics at the Dublin Institute for Advanced Studies, for stimulating discussions of great assistance to me.

J. L. S.

DUBLIN
9 *December* 1952

CHAPTER I

INTRODUCTION

This book is intended as a contribution to 'mathematical' physics rather than to 'naturalistic' physics, as I have elsewhere interpreted those words (Synge, 1952).* Briefly this means that the creation of a coherent mathematical theory, with assumptions clearly set forth, is the primary object, physical interpretation in terms of experiment being relegated to a secondary position. Nevertheless, the theory has a close connexion with physical reality, and its scope and purpose are best explained by comparing optics and mechanics.

In optics there exist two, and I think only two, coherent mathematical theories; these we may conveniently call Hamilton's theory and Maxwell's theory, although it might be better to say the d'Alembert-Maxwell theory, since the useful and coherent theory based on the scalar wave equation should of course be included (cf. Baker and Copson, 1950).

Hamilton's theory starts from a variational principle $\delta \int v \, ds = 0$, where v is a medium-function or index of refraction, depending on position and direction; and from this principle the theory constructs the properties of systems of rays and of the waves associated with them (extremals and transversals, in the language of the calculus of variations). We need not think of anything as moving; the theory may be regarded as a statical one, the rays being fixed curves in space and the waves fixed surfaces. Neither wave-length nor frequency is involved; the waves form a continuous set of surfaces, not distinguished as crests and troughs. This theory, whether in the form preferred by Hamilton or otherwise, has been the subject of many books under the general title 'geometrical optics' and is the basic theory used in the design of optical instruments, even if the designer should use no more than the law of refraction for a ray, since that law may be regarded as a deduction from the theory.

Maxwell's theory starts from a set of partial differential equations, and the solution of any optical problem involves the solution of these

* See references at the end of the book, p. 164.

equations with assigned boundary conditions. Once the problem has been clearly stated in these terms, the difficulties of solution are purely mathematical. But these difficulties are great, and so the number of exact solutions of Maxwell's equations remains comparatively small.

The great body of optics (we may call it 'physical optics') lies between the theories of Hamilton and Maxwell. Hamilton's theory fails physically when the wave-lengths involved cannot be neglected, whereas Maxwell's theory is too elaborate and difficult in its exact form. Physical optics takes Hamilton's theory and extends it by adding the concepts of wave-length, secondary waves and interference, justifying this procedure by semi-intuitive appeals to Maxwell's theory; by this I mean that complete boundary-value problems are rarely solved—no one would think of doing so in investigating the diffraction pattern at the focus of a telescope. Physical optics, so understood, is essential to physics, being used all the time by physicists who can find out what they want only by means of it. It is not a coherent mathematical structure, and it would be useless to try to make it one.

Thus we have in optics two coherent mathematical theories and, in between them, a semi-coherent but very useful body of extensions and approximations, and (one might add) outside them a semi-coherent body of theory dealing with the photon.

As for mechanics, we see in the Newtonian theory of the motion of a particle the analogue of geometrical optics and in Schrödinger's wave mechanics the analogue of Maxwell's theory. All these are coherent mathematical theories. But the analogy between Newtonian mechanics and geometrical optics is completed only when we supplement the former by thinking of waves in association with the paths of particles. This completion was actually present in Hamilton's theory, since he made it so wide as to include both corpuscular and wave theories of light, and in the former interpretation his surfaces of constant action are the waves in question. Thus, since the time of Hamilton we have actually had what might be called 'Newtonian geometrical mechanics', based on the principle of Maupertuis, $\delta \int v \, ds = 0$, where v is given in terms of energy by $\frac{1}{2}mv^2 = E - V$. Furthermore, since the ray-wave relationship is merely part of the mathematical calculus of variations, it is available as soon as we write down a variational principle of this form. This can be done for a limited but important class of relativistic problems in which there is an integral of energy, and so we recognize (subject to this limitation) the existence of 'relativistic geometrical mechanics'.

In making his great contribution to physics, de Broglie (de Broglie and Brillouin, 1928, p. 55) was not concerned to analyse his work on these lines; he preferred a more suggestive physical approach. But when we look back at it, we see that the de Broglie waves had been in (mathematical) existence since the time of Hamilton. What de Broglie did was to add those things that change geometrical optics into physical optics, namely, wave-length, secondary waves and interference, so creating what we may call 'physical mechanics', bearing to Schrödinger's wave mechanics the relationship that physical optics bears to Maxwell's theory. Pursuing the analogy, we may expect the de Broglie waves to constitute a permanent part of the physics of the future, offering easy solutions of problems in cases where the Schrödinger wave equation is too difficult to apply.

If that is so, we should try to make the underlying coherent mathematical theory as simple and general as possible. 'Geometrical mechanics' (the term is de Broglie's) should be as mathematically precise as geometrical optics. As indicated above, Hamilton completed that task for Newtonian mechanics, but the theory of 'relativistic geometrical mechanics' remains incomplete. The purpose of this book is to complete it, at least as regards essentials.

The task is very much easier than might be supposed, for all we have to do is to apply Hamilton's formal methods of geometrical optics in the space-time of Minkowski, making allowance for its four-dimensionality and the indefinite character of its metric. Where Hamilton saw a source of light emitting rays, we see an event in space-time with world lines drawn from it into the future. These we may interpret physically as the possible histories of a particle in whose history one event is accurately known, and whose velocity is therefore quite unknown, according to the Heisenberg uncertainty principle. Where Hamilton saw a wave (fixed two-dimensional surface), we see a three-dimensional surface in space-time; this represents the history of a two-dimensional surface moving in the observer's space—it is, in fact, a de Broglie wave moving with a velocity which we can calculate from the velocity of the ray. The group velocity of the waves is the particle (or ray) velocity, and this is not a question of the interference of waves with different frequencies, for the theory here does not include the concept of frequency; it is a consequence of the primitive definition of group velocity as the velocity of a region of disturbance on the boundaries of which waves disappear and are created.

Chapters II and III are devoted to this pure geometrical mechanics. To pass to physical mechanics, we must add frequency and wavelength (or some relativistically invariant equivalent), and this addition is to be regarded essentially as a process of quantization. This is discussed in Chapter IV, where, after some preliminary work with the second-order wave equation, a primitive quantization is defined: adjacent 3-waves of equal phase are separated by a quantum h of action, measured along the rays in space-time. It is rather surprising that when we discuss the quantization of a hydrogenic atom according to this simple plan we obtain precisely the usual fine-structure formula; the simple Zeeman effect also comes out.

But one should realize that Chapters II and III constitute in themselves a coherent theory of the geometrical mechanics of a particle, quite independent of the quantization of Chapter IV, being in fact no more than the systematic development of certain mathematical properties of sets of extremals and transversals in Minkowskian space-time.

The whole theory may be regarded as a synthesis of the ideas of Hamilton, Minkowski and de Broglie. The method used is that of Hamilton's optics (Hamilton, 1931, in particular pp. 168–71) and not his much better known dynamical method (Hamilton, 1941, in particular pp. 164–8). The essential difference is that in the optical method the coordinates are given equal rights (the three coordinates of space for Hamilton, the four coordinates of space-time for us), whereas in the dynamical method the time t is privileged over the other coordinates. That is of course natural and proper in Newtonian physics, but to a relativitist the dynamical method cannot but be something of a monstrosity, offering a clumsy approach to space-time problems, as unsatisfying as an approach to the geometry of surfaces through an equation of the form $z = f(x, y)$ rather than $F(x, y, z) = 0$. There is a good analogy here, the difference between the dynamical and optical methods being as trivial or as important as the difference between $z = f(x, y)$ and $F(x, y, z) = 0$; in the dynamical method we express the energy in terms of the other variables, obtaining a Hamiltonian H, whereas in the optical method (applied to mechanics) we let the energy take its place beside the other variables in a slowness equation $\Omega = 0$. It is true that we have only to solve $F(x, y, z) = 0$ for z in order to get $z = f(x, y)$, but there would be little beauty or power in the theory of quadric surfaces, for example, if we had to use the equation

$$z = c^{-1} \{ -gx - fy \pm [(gx + fy)^2 - c(ax^2 + 2hxy + by^2 - 1)]^{\frac{1}{2}} \}$$

instead of

$$ax^2 + by^2 + cz^2 + 2fyz + 2gzx + 2hxy - 1 = 0.$$

The connexion between the two methods is discussed in § 5·5.

In order to understand the present theory properly, it is absolutely essential to use both *geometrical* representation (pictures of space-time in the style of Minkowski) and the more ordinary *kinematical* representation in which one sees points and surfaces moving in the space of a Galilean observer.

In Chapter V the Hamiltonian method is exhibited in a more general form and the theory applied to problems involving two particles.

To illustrate the rather abstract theory of Chapter II, simple examples are inserted without regard to their possible physical meaning. These examples are distinguished from the main body of the argument by the fact that the equations for the examples are not numbered.

GENERAL THEORY OF RAYS AND WAVES IN SPACE-TIME

2.1. Rays in space-time

We think of flat Minkowskian space-time with rectangular Cartesian coordinates x_1, x_2, x_3 and an imaginary time coordinate $x_4 = ict$, c being the velocity of light. We denote these coordinates briefly by x_r, small Latin suffixes having here and throughout the book the range of values 1, 2, 3, 4, with summation understood for a repeated suffix. Later we shall use Greek suffixes for the range 1, 2, 3, again with the summation convention.

Consider an event x_r and all time-like directions there, pointing into the future; any such direction corresponds to a unit 4-vector α_r such that
$$\alpha_r \alpha_r = -1, \quad \alpha_4/i > 0. \tag{2.1.1}$$
Let $f(x, \alpha)$ be an invariant function of position and direction in space-time. Following Hamilton (1931, p. 169) we shall call f the *medium-function*, but we shall use the symbol f instead of Hamilton's v, since the latter will be needed for velocity. We shall take f to be positive. We require f to be defined only for time-like directions α_r; we are not interested in space-like directions or null directions.

By virtue of (2.1.1) we can make f positive homogeneous of degree unity in the α's, and this we shall always suppose done when we have occasion to take partial derivatives. Then
$$f(x, k\alpha) = kf(x, \alpha) \quad (k > 0), \qquad \alpha_r \frac{\partial f}{\partial \alpha_r} = f. \tag{2.1.2}$$
Thus, for example, if we were given $f = 1$, we would secure the required homogeneity by writing $f = (-\alpha_r \alpha_r)^{\frac{1}{2}}$; similarly, we would modify $f = g_{rs}(x) \alpha_r \alpha_s$ by writing it $f = g_{rs}(x) \alpha_r \alpha_s (-\alpha_n \alpha_n)^{-\frac{1}{2}}$.

Rays in space-time are curves satisfying the variational principle
$$\delta \int f(x, \alpha) \, ds = 0 \tag{2.1.3}$$
for fixed end-events, ds being the Minkowskian element, so that $ds^2 = -dx_r \, dx_r$, the minus sign occurring since we confine our

attention to time-like curves. Here $\alpha_r = dx_r/ds$ the velocity 4-vector. By (2.1.3) rays satisfy the Euler-Lagrange equations

$$\frac{d}{ds}\frac{\partial f}{\partial \alpha_r} - \frac{\partial f}{\partial x_r} = 0. \qquad (2.1.4)$$

In general, a ray is determined by an initial event and initial direction in space-time, or by two events $P'(x'), P(x)$ on it. Following the notation of Hamilton, we shall always use P' for the initial event and P for the final event, so that the direction of $P'P$ points into the future.

We define the *action* along any time-like world line to be

$$\int f(x, \alpha) \, ds$$

taken along it, and so the variational principle (2.1.3) may be called the *law of stationary action*. We define the *characteristic function* $V(x', x)$ or $V(P', P)$ to be the action along the ray joining $P'(x')$ and $P(x)$, that is,

$$V(x', x) = \int_{P'}^{P} f(x, \alpha) \, ds, \qquad (2.1.5)$$

the integral being taken along the ray.

On varying the end-events, we find

$$\frac{\partial V}{\partial x_r} = \frac{\partial f}{\partial \alpha_r}, \quad \frac{\partial V}{\partial x_r'} = -\frac{\partial f'}{\partial \alpha_r'}, \qquad (2.1.6)$$

where $f' = f(x', \alpha')$. Since $\partial f/\partial \alpha_r$ is homogeneous of degree zero in the α's, it is a function only of their three ratios (and, of course, of the x's). These three ratios can be eliminated from the four equations standing on the left in (2.1.6), and a similar elimination may be performed on the other four equations. Hence we get *Hamilton's two partial differential equations*

$$\Omega\left(-\frac{\partial V}{\partial x}, x\right) = 0, \quad \Omega\left(\frac{\partial V}{\partial x'}, x'\right) = 0. \qquad (2.1.7)$$

We may call the first of these the *Hamilton-Jacobi equation* (but not the second, since Jacobi spurned its analogue in dynamics as a mere nuisance!). If we differentiate the Hamilton-Jacobi equation with respect to x_s' and then eliminate the derivatives of Ω, we obtain *Hamilton's determinantal equation*

$$\det \frac{\partial^2 V}{\partial x_r \, \partial x_s'} = 0. \qquad (2.1.8)$$

As an illustrative example, take $f(x, \alpha) = (-\alpha_r \alpha_r)^{\frac{1}{2}}$. It follows from (2.1.4) that the rays are straight lines in space-time, and hence by (2.1.5)

$$V(x', x) = [-(x_r - x'_r)(x_r - x'_r)]^{\frac{1}{2}}.$$

To obtain the Hamilton-Jacobi equation, we have by (2.1.6)

$$\frac{\partial V}{\partial x_r} = -\alpha_r(-\alpha_n \alpha_n)^{-\frac{1}{2}},$$

and so

$$\frac{\partial V}{\partial x_r}\frac{\partial V}{\partial x_r} = -1, \quad \Omega = \frac{\partial V}{\partial x_r}\frac{\partial V}{\partial x_r} + 1 = 0.$$

We have further

$$\frac{\partial V}{\partial x_r} = -\frac{x_r - x'_r}{V}, \quad \frac{\partial V}{\partial x'_s} = \frac{x_s - x'_s}{V},$$

$$\frac{\partial^2 V}{\partial x_r \partial x'_s} = \frac{\delta_{rs}}{V} + \frac{x_r - x'_r}{V^2}\frac{\partial V}{\partial x'_s} = \frac{\delta_{rs}}{V} + \frac{(x_r - x'_r)(x_s - x'_s)}{V^3}.$$

Hence

$$\frac{\partial^2 V}{\partial x_r \partial x'_s}(x_s - x'_s) = \frac{x_r - x'_r}{V} - \frac{(x_r - x'_r)V^2}{V^3} = 0,$$

and so we verify (2.1.8).

We define the *slowness 4-vector* σ_r by

$$\sigma_r = -\frac{\partial f}{\partial \alpha_r}. \tag{2.1.9}$$

Hence we have by (2.1.2) the important relation

$$\sigma_r \alpha_r = -f. \tag{2.1.10}$$

By (2.1.6) the variation of $V(x', x)$, resulting from varying the end-events, may be expressed in terms of the slowness 4-vectors at the initial and final events in the form

$$\delta V = -\sigma_r \delta x_r + \sigma'_r \delta x'_r. \tag{2.1.11}$$

We note that

$$\frac{\partial V}{\partial x_r} = -\sigma_r, \quad \frac{\partial V}{\partial x'_r} = \sigma'_r. \tag{2.1.12}$$

In writing the definition (2.1.9), I have inserted a minus sign where Hamilton used a plus sign in his optics. It is convenient to make this change for a relativistic theory. Since f is positive, (2.1.10) makes $\sigma_r \alpha_r$ negative. Since α_r is time-like and points into the future, this makes σ_r and α_r lie on the same side (the future side) of the 3-plane $\Pi(\alpha)$ normal to α_r, and this is the most convenient way to have it. Three different cases may arise:

$$
\left.
\begin{array}{lll}
(a) & \sigma_r \text{ time-like:} & \sigma_r\sigma_r < 0, \\
(b) & \sigma_r \text{ null:} & \sigma_r\sigma_r = 0, \\
(c) & \sigma_r \text{ space-like:} & \sigma_r\sigma_r > 0.
\end{array}
\right\}
\qquad (2.1.13)
$$

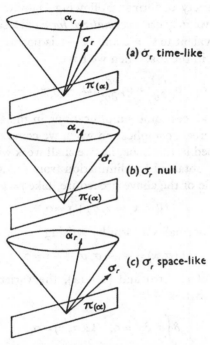

Fig. 2.1. Space-time diagrams showing the null-cone and three possible positions of the slowness 4-vector σ_r: (a) σ_r time-like, (b) σ_r null, (c) σ_r space-like.

These three cases are shown in fig. 2.1. We note that in cases (a) and (b) every Galilean observer finds σ_4/i positive; in case (c) he may find σ_4/i positive or negative according to his choice of frame of reference, i.e. he may find the vector σ_r pointing into his future or into his past.

In the example cited above, $f = (-\alpha_r\alpha_r)^{\frac{1}{2}}$, we have

$$
\sigma_r = \alpha_r(-\alpha_n\alpha_n)^{-\frac{1}{2}} = \alpha_r, \qquad \sigma_r\sigma_r = -1,
$$

and so this is case (a).

If we eliminate the α's from (2.1.9), we get the *slowness equation*

$$\Omega(\sigma, x) = 0, \qquad (2.1.14)$$

the same functional form as in (2.1.7). This relationship between the σ's and the x's may be written in a variety of forms, just as in ordinary geometry the equation of a surface $F(x, y, z) = 0$ may be written in a variety of forms. Following Hamilton, except for a change of sign, we may use a *normalized form* of Ω, say Ω', such that (2.1.14) is equivalent to $\Omega' = 0$ and $\Omega' - 1$ is positive homogeneous of degree unity in the σ's. Then we have

$$\Omega'(\sigma, x) = 0, \qquad \sigma_r \frac{\partial \Omega'}{\partial \sigma_r} = \sigma_r \frac{\partial}{\partial \sigma_r}(\Omega' - 1) = \Omega' - 1 = -1. \quad (2.1.15)$$

This device produces some simplifications in the abstract theory, but it tends to create complication when we come to applications; it will not be used in this book, i.e. we shall work with any form of the function Ω, obtained by elimination from (2.1.9).

As an example of the above procedure, take $f = (-\alpha_r \alpha_r)^{\frac{1}{2}}$. Then

$$\Omega(\sigma, x) = \sigma_r \sigma_r + 1 = 0,$$

and we get the normalized form by writing

$$\Omega'(\sigma, x) = -(-\sigma_r \sigma_r)^{\frac{1}{2}} + 1 = 0.$$

On account of (2.1.10) and (2.1.14), the variational equation (2.1.3) may be written

$$\delta \int \sigma_r \, dx_r = 0, \qquad \Omega(\sigma, x) = 0, \qquad (2.1.16)$$

the vector field σ_r being regarded as arbitrary except for the last condition. Since $\delta x_r = 0$ at the end-events, we have then

$$0 = \int (\delta \sigma_r \, dx_r + \sigma_r \, \delta dx_r) = \int (\delta \sigma_r \, dx_r - \delta x_r \, d\sigma_r)$$

for all variations subject to

$$\frac{\partial \Omega}{\partial \sigma_r} \delta \sigma_r + \frac{\partial \Omega}{\partial x_r} \delta x_r = 0.$$

Hence on a ray we have

$$dx_r = dw \frac{\partial \Omega}{\partial \sigma_r}, \qquad d\sigma_r = -dw \frac{\partial \Omega}{\partial x_r},$$

where dw is an infinitesimal multiplier. Thus we have the *Hamiltonian form of the equations of a ray*

$$\frac{dx_r}{dw} = \frac{\partial\Omega}{\partial\sigma_r}, \quad \frac{d\sigma_r}{dw} = -\frac{\partial\Omega}{\partial x_r}. \qquad (2.1.17)$$

Here w is a parameter which, as we see by making use of (2.1.10), is related to the Minkowskian arc length s by

$$f\frac{ds}{dw} = -\sigma_r\frac{\partial\Omega}{\partial\sigma_r}. \qquad (2.1.18)$$

If we use the normalized Ω', we get dw = dV = fds.

Thus, for the example $f = (-\alpha_r\alpha_r)^{\frac{1}{2}}$, we have the slowness equation
$$\Omega(\sigma, x) = \sigma_r\sigma_r + 1 = 0;$$

the Hamiltonian equations (2.1.17) read

$$\frac{dx_r}{dw} = 2\sigma_r, \quad \frac{d\sigma_r}{dw} = 0,$$

and (2.1.18) gives

$$f\frac{ds}{dw} = -2\sigma_r\sigma_r = 2, \quad dw = \tfrac{1}{2}fds.$$

The unit 3-wave and the 3-surface of slowness. The function $f(x, \alpha)$ and the equation $\Omega(\sigma, x) = 0$ can be interpreted geometrically in space-time, and there is an important relation between them. Let $P(x)$ be any event and let y_r be the coordinates of any other event relative to P. Consider two 3-spaces S and W defined by the equations

$$S: \quad f(x, y) = 1, \qquad (2.1.19)$$
$$W: \quad \Omega(y, x) = 0. \qquad (2.1.20)$$

Note that in these equations x stands for fixed numbers x_r and y for current coordinates y_r. We call S the *unit 3-wave* at P and W the *3-surface of slowness* at P (cf. Hamilton, 1931, p. 291).

To explore the relationship between S and W, consider the two unit pseudospheres $U(\epsilon)$ with centre P; their equations are $y_ry_r = \epsilon$ ($\epsilon = \pm 1$). The polar 3-planes of an event y_r' with respect to them are respectively

$$y_ry_r' = \epsilon. \qquad (2.1.21)$$

If we let y_r' range over S, the polar reciprocal of S with respect to $U(\epsilon)$ is the envelope of the 3-planes (2.1.21), and the event of contact y_r of the 3-plane (2.1.21) with this envelope is to be found by solving (2.1.21) and the four equations

$$\theta y_r = \frac{\partial f(x, y')}{\partial y'_r}, \qquad (2.1.22)$$

θ being a factor of proportionality. Multiplying this by y'_r and using (2.1.21) and the homogeneity of f, we get $\theta = \epsilon f(x, y') = \epsilon$, since y'_r lies on S. Elimination of y'_r from (2.1.22) then gives

$$\Omega(-\epsilon y, x) = 0 \qquad (2.1.23)$$

(cf. (2.1.14)). But this is the equation (2.1.20) of W, provided that $\epsilon = -1$; and so we have this result: *the 3-surface of slowness W is the*

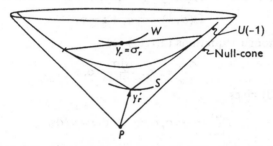

Fig. 2.2. Space-time diagram showing the reciprocal relationship between the unit 3-wave S and the 3-surface of slowness W, drawn for the case where σ_r is time-like.

polar reciprocal of the unit 3-wave S (and of course conversely) with respect to the unit pseudosphere $y_r y_r = -1$, that is, the unit pseudosphere lying inside the null-cone. Fig. 2.2 shows the relationship.

For the example $f(x, \alpha) = (-\alpha_r \alpha_r)^{\frac{1}{2}}$, the unit 3-wave S is $(-y_r y_r)^{\frac{1}{2}} = 1$ or $y_r y_r = -1$, or explicitly

$$y_1^2 + y_2^2 + y_3^2 + y_4^2 = -1.$$

This is itself the unit pseudosphere. The 3-surface of slowness W is $y_r y_r + 1 = 0$, the same unit pseudosphere. So we get the very simple case in which S and W both coincide with the unit pseudosphere and each is its own reciprocal.

If we take the slightly more general example,

$$f(x, \alpha) = k(-\alpha_r \alpha_r)^{\frac{1}{2}},$$

where k is a constant, we have for S the equation $y_r y_r = -k^{-2}$ and for W the equation $y_r y_r = -k^2$. Now we have two concentric pseudospheres and the product of their radii is unity, just as the product of the radii of two concentric spheres is unity if one is the polar reciprocal of the other with respect to the unit sphere concentric with them.

This reciprocal relationship between S and W is well known in its general form in the calculus of variations. It is of fundamental importance in the present theory because it contains the relationship between the ray and wave velocities in the de Broglie theory, although when we come to that we shall use a direct method. It might also be used to throw geometrical light on the analytic argument which follows immediately.

Determination of f from Ω. Usually we suppose the medium-function $f(x, \alpha)$ given, and from it we derive by elimination the equation $\Omega(\sigma, x) = 0$. Now let us see how to obtain f from a given slowness equation

$$\Omega(\sigma, x) = 0. \tag{2.1.24}$$

We regard σ_r, x_r as eight variables bound only by this equation. We think then of four more variables α_r connected by

$$\alpha_r \alpha_r = -1. \tag{2.1.25}$$

We seek to define the α's as functions of the σ's and x's and also to define a positive function $f(x, \alpha)$, positive homogeneous of degree unity in the α's and satisfying the relation (2.1.10):

$$\sigma_r \alpha_r = -f(x, \alpha). \tag{2.1.26}$$

Regarding the twelve variables (σ, x, α) as arbitrary and independent except for the three equations written above, we get, on taking variations,

$$\frac{\partial \Omega}{\partial \sigma_r} \delta \sigma_r + \frac{\partial \Omega}{\partial x_r} \delta x_r = 0, \tag{2.1.27}$$

$$\alpha_r \delta \alpha_r = 0, \tag{2.1.28}$$

$$\sigma_r \delta \alpha_r + \alpha_r \delta \sigma_r + \frac{\partial f}{\partial \alpha_r} \delta \alpha_r + \frac{\partial f}{\partial x_r} \delta x_r = 0. \tag{2.1.29}$$

Now (2.1.29) is to hold whenever (2.1.27) and (2.1.28) hold, and so we have (with Lagrange multipliers θ and ϕ)

$$\alpha_r = \theta \frac{\partial \Omega}{\partial \sigma_r}, \quad \sigma_r + \frac{\partial f}{\partial \alpha_r} = \phi \alpha_r, \quad \frac{\partial f}{\partial x_r} = \theta \frac{\partial \Omega}{\partial x_r}. \tag{2.1.30}$$

If we multiply the second of these by α_r and use (2.1.26) and the required homogeneity of f, we get $\phi = 0$ and hence

$$\sigma_r = -\frac{\partial f}{\partial \alpha_r}, \qquad (2.1.31)$$

as in (2.1.9), where it was a definition, but now (starting from $\Omega = 0$) a deduction. If we multiply the first of (2.1.30) by $-\sigma_r$, we get

$$f = -\sigma_r \alpha_r = -\theta \sigma_r \frac{\partial \Omega}{\partial \sigma_r}. \qquad (2.1.32)$$

Let us now collect together (2.1.24), the first of (2.1.30), and (2.1.32):

$$\Omega(\sigma, x) = 0, \quad \alpha_r = \theta \frac{\partial \Omega}{\partial \sigma_r}, \quad f = -\theta \sigma_r \frac{\partial \Omega}{\partial \sigma_r}. \qquad (2.1.33)$$

Here are six equations. From them we can eliminate the five quantities σ_r, θ, obtaining a relation between f, α_r, x_r. Solving this for f, we get the required $f(x, \alpha)$. That it is homogeneous of degree unity in the α's is indicated by the form of the equations, since if they are satisfied by f, α_r, θ, they are also satisfied by kf, $k\alpha_r$, $k\theta$. However, on carrying out the algebraic calculations, we may find the function $f(x, \alpha)$ to be many-valued, and the effect of multiplying by a negative k may be to switch the branches. All we need is *positive* homogeneity, and that is assured by the form of (2.1.33) if we pick the appropriate branch.

For example, if $\Omega(\sigma, x) = \frac{1}{2}(\sigma_r \sigma_r + 1) = 0$, then (2.1.33) will read

$$\sigma_r \sigma_r + 1 = 0, \quad \alpha_r = \theta \sigma_r, \quad f = -\theta \sigma_r \sigma_r,$$

and so we get

$$f = \theta, \quad \alpha_r \alpha_r = \theta^2 \sigma_r \sigma_r = -\theta^2, \quad \theta = \pm(-\alpha_r \alpha_r)^{\frac{1}{2}}.$$

Thus $f(x, \alpha) = \pm(-\alpha_r \alpha_r)^{\frac{1}{2}}$. We select the positive root, and this is positive homogeneous, but not homogeneous with respect to a negative multiplier.

Let us consider two further examples of the transitions from f to Ω and from Ω to f. These are of some interest in connexion with the general theory of relativity.

Consider first
$$f(x, \alpha) = [-g_{rs}(x)\alpha_r \alpha_s]^{\frac{1}{2}},$$

where $g_{rs} = g_{sr}$. Then

$$\sigma_r = -\frac{\partial f}{\partial \alpha_r} = g_{rs}\frac{\alpha_s}{f}.$$

Let g^{mn} be the conjugate of g_{mn}, defined by $g^{rp}g_{rq}=\delta^p_q$. Then we have

$$g^{pr}\,\sigma_r=\alpha_p/f,$$

$$f^2=-g_{pq}\alpha_p\alpha_q=-g_{pq}g^{pr}\,\sigma_r g^{qs}\,\sigma_s f^2=-g^{rs}\,\sigma_r\sigma_s f^2,$$

and so we get $\Omega(\sigma,x)=g^{rs}(x)\,\sigma_r\sigma_s+1=0.$

As a second example, take the slowness equation

$$\Omega(\sigma,x)=\tfrac12 g^{rs}(x)\,[\sigma_r-A_r(x)]\,[\sigma_s-A_s(x)]+\tfrac12=0.$$

Then $\dfrac{\partial\Omega}{\partial\sigma_r}=g^{rs}(\sigma_s-A_s),\quad \sigma_r\dfrac{\partial\Omega}{\partial\sigma_r}=2\Omega-1+g^{rs}A_r(\sigma_s-A_s)$

$$=g^{rs}A_r(\sigma_s-A_s)-1,$$

and (2.1.33) gives

$$\Omega=0,\quad \alpha_r=\theta g^{rs}(\sigma_s-A_s),\quad f=-\theta[g^{rs}A_r(\sigma_s-A_s)-1].$$

Hence $\sigma_s-A_s=\theta^{-1}g_{rs}\alpha_r,$

$$2\Omega=\theta^{-2}g^{rs}g_{rp}\alpha_p g_{sq}\alpha_q+1=\theta^{-2}g_{pq}\alpha_p\alpha_q+1=0,$$

$$\theta^2=-g_{pq}\alpha_p\alpha_q,$$

and so the medium-function is

$$f(x,\alpha)=-g^{rs}A_r g_{sp}\alpha_p+\theta=-A_r\alpha_r+(-g_{pq}\alpha_p\alpha_q)^{\frac12}.$$

Hamilton's dynamical method. If frequent appeals to the geometry of Minkowskian space-time and the symmetric treatment of the four coordinates x_r are to anyone a hindrance rather than a help, it is still open to him to follow Hamilton's dynamical method. For the invariant element of action may be written in terms of a Lagrangian function as*

$$f(x_1,x_2,x_3,x_4,\alpha_1,\alpha_2,\alpha_3,\alpha_4)\,\mathrm{d}s=-L(x_1,x_2,x_3,t,\dot x_1,\dot x_2,\dot x_3)\,\mathrm{d}t,$$

$$(2.1.34)$$

where $\dot x_\rho=\mathrm{d}x_\rho/\mathrm{d}t$, Greek suffixes having the range 1, 2, 3. Then the variational principle (2.1.3) is equivalent to

$$\delta\int L(x,t,\dot x)\,\mathrm{d}t=0,\qquad (2.1.35)$$

the terminal values of x_ρ and t being held fixed. Hence we obtain for the rays the usual Lagrangian equations

$$\frac{\mathrm{d}}{\mathrm{d}t}\frac{\partial L}{\partial\dot x_\rho}-\frac{\partial L}{\partial x_\rho}=0,\qquad (2.1.36)$$

equivalent, of course, to (2.1.4).

* The minus sign is unimportant, but convenient in application to (3.3.1.).

If we introduce p_ρ by $p_\rho = \partial L/\partial \dot{x}_\rho$, we can write a Hamiltonian

$$H(p,x,t) = \dot{x}_\rho p_\rho - L(x,t,\dot{x}), \qquad (2.1.37)$$

and express the equations of the rays in the form

$$\dot{x}_\rho = \frac{\partial H}{\partial p_\rho}, \quad \dot{p}_\rho = -\frac{\partial H}{\partial x_\rho}, \qquad (2.1.38)$$

instead of (2.1.17). The Hamilton-Jacobi equation is

$$\frac{\partial A}{\partial t} + H\left(\frac{\partial A}{\partial x}, x, t\right) = 0 \qquad (2.1.39)$$

instead of (2.1.7), where $A = \int L\,dt = -\int f\,ds = -V$.

To mix the two methods would be confusing, and, while it would be interesting to carry through the argument in dynamical form, the mathematical structure of the optical method is simpler and it will be used throughout. A further discussion of the connexion between the two methods will be found in § 5.5.

2.2. Waves in space-time

From an event $P'(x')$, which we shall call a *source-event*, let us draw all possible rays into the future. Then the locus in space-time obtained by taking on these rays those events $P(x)$ for which the action $\int_{P'}^{P} f\,ds$ has a constant value is a 3-space Σ; we shall call it a *3-wave*. Thus the source-event P' determines a singly infinite set of 3-waves (fig. 2.3) with equations

$$V(P',P) = \text{const.}, \qquad (2.2.1)$$

V being the characteristic function.

For example, if $f = (-\alpha_r \alpha_r)^{\frac{1}{2}}$, then

$$V(P',P) = [-(x_r - x_r')(x_r - x_r')]^{\frac{1}{2}},$$

and the 3-waves consist of the pseudospheres of all radii having P' for centre and lying in the future null-cone with vertex at P'.

The above definition of a 3-wave is included in the following more general definition. We take a 3-space Σ_0, arbitrary except for the limitation that its normal must lie in the domain of possible directions of σ_r satisfying $\Omega = 0$. Thus, for example, if the function $f(x, \alpha)$ is such that

$$\frac{\partial f}{\partial \alpha_r}\frac{\partial f}{\partial \alpha_r} < 0, \qquad (2.2.2)$$

then by $(2.1.9)$ σ_r is time-like, and so we must restrict ourselves in this case to 3-spaces Σ_0 with time-like normals.

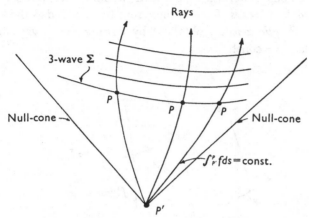

Fig. 2.3. Space-time diagram of 3-waves from a source-event P'.

With this understanding, then, there exists a σ_r at P' (say σ_r') which is normal to Σ_0 and satisfies

$$\Omega(\sigma', x') = 0. \qquad (2.2.3)$$

The relations

$$\frac{\partial f'}{\partial \alpha_r'} = -\sigma_r' \qquad (2.2.4)$$

then determine a direction α_r' at P'. Taking these directions at all events on Σ_0 and drawing the rays in these directions (by solving the differential equations $(2.1.4)$, or equivalently $(2.1.17)$), we generate a congruence of rays. If we measure off any constant value of the action $\int_{P'}^{P} f\,ds$ along these rays, the events P so obtained generate a 3-space Σ which we shall call a 3-wave (fig. 2.4). If we take $\int_{P'}^{P} f\,ds = 0$, Σ collapses on Σ_0.

To cite the simplest possible example, take $f = (-\alpha_r \alpha_r)^{\frac{1}{2}}$. Then $(2.2.4)$ gives $\sigma_r' = \alpha_r'$, and so the ray direction α_r' is normal to Σ_0, since σ_r' is normal to Σ_0. In this case each 3-wave is obtained from

Σ_0 by measuring off equal Minkowskian distances along the normals of Σ_0. Let it be noted, however, that these properties are not general. In general, the rays are *not* normal to the 3-waves, nor are two 3-waves separated by constant Minkowskian distances measured along the rays.

For reasons which will appear later, we may refer to the 3-waves as *de Broglie* 3-*waves*. Each of them may be regarded as the history of a *de Broglie* 2-*wave*, obtained by cutting the 3-wave by the hyperplane $t = \text{const.}$

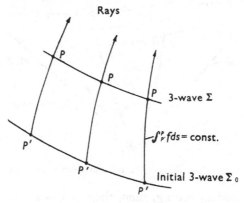

Fig. 2.4. Space-time diagram of 3-wave Σ generated by initial 3-wave Σ_0.

The more general definition of a 3-wave (fig. 2.4) includes as a special case the previous definition (fig. 2.3), the initial 3-wave Σ_0 degenerating to a single source-event. In the more general definition, the 3-waves are again given in terms of the characteristic function by the equation

$$V(P', P) = \text{const.,} \qquad (2.2.5)$$

but now P' is not fixed; it ranges over Σ_0, being always connected with P by the condition that the ray $P'P$ has at P' a slowness 4-vector normal to Σ_0. Thus in (2.2.5) V is to be regarded as depending on P only.

The set of 3-waves generated in this way from a 3-space Σ_0 has an important *group property*, the same set of 3-waves being generated from any one of them. To see this, we recall (2.1.11), which tells us that, for any variations of the end-events of a ray,

$$\delta V = -\sigma_r \delta x_r + \sigma_r' \delta x_r'. \qquad (2.2.6)$$

Now if we vary P' on Σ_0 and P on Σ, we have $\delta V = 0$ and $\sigma_r' \delta x_r' = 0$; hence $\sigma_r \delta x_r = 0$, showing that the slowness 4-vector at P is normal to Σ, and consequently the ray $P'P$ has the correct direction for carrying on the process, starting from Σ.

2.3. Gauge transformations

If we are given a medium-function $f(x, \alpha)$, this function determines the rays by $\delta \int f \, ds = 0$, and everything else may be worked out in terms of it.

If we change from a medium-function $f(x, \alpha)$ to a new medium-function $f^*(x, \alpha)$, then in general the rays become completely different. But consider the case where

$$f^*(x, \alpha) = f(x, \alpha) - \frac{\partial \phi}{\partial x_r} \alpha_r, \qquad (2.3.1)$$

where ϕ is some given function of the coordinates x_r. Then, for any curve in space-time,

$$\int_{P'}^{P} f^* \, ds = \int_{P'}^{P} f \, ds - \int_{P'}^{P} \frac{\partial \phi}{\partial x_r} \, dx_r$$

$$= \int_{P'}^{P} f \, ds + \phi(x') - \phi(x). \qquad (2.3.2)$$

On varying the curve, keeping the end-events fixed, we get

$$\delta \int_{P'}^{P} f^* \, ds = \delta \int_{P'}^{P} f \, ds, \qquad (2.3.3)$$

and so the variational equation $\delta \int f^*(x, \alpha) \, ds = 0$ gives precisely the same rays as does $\delta \int f(x, \alpha) \, ds = 0$.

We call (2.3.1) a *gauge transformation* of the medium-function. Then we know that *rays are invariant under a gauge transformation*. The important thing is that the rest of the theory is not gauge-invariant. This means, for example, that if we take the rays from a source-event, the 3-waves associated with them are changed by a gauge transformation. Or if we start from a given initial 3-wave Σ_0, the directions of the rays coming from it will be changed by a gauge transformation.

Let us put down here for reference the changes induced by a

gauge transformation, the unstarred symbols referring to theory based on $f(x, \alpha)$ and the starred symbols to that based on $f^*(x, \alpha)$, connected with $f(x, \alpha)$ by (2.3.1):

Medium-function

$$f(x, \alpha) \qquad\qquad f^*(x, \alpha) = f(x, \alpha) - \frac{\partial \phi}{\partial x_r}\alpha_r$$

Characteristic function

$$V(x', x) \qquad\qquad V^*(x', x) = V(x', x) + \phi(x') - \phi(x)$$

$$\frac{\partial V}{\partial x_r} = \frac{\partial f}{\partial \alpha_r} \qquad\qquad \frac{\partial V^*}{\partial x_r} = \frac{\partial f^*}{\partial \alpha_r} = \frac{\partial f}{\partial \alpha_r} - \frac{\partial \phi}{\partial x_r}$$

$$\frac{\partial V}{\partial x_r'} = -\frac{\partial f'}{\partial \alpha_r'} \qquad\qquad \frac{\partial V^*}{\partial x_r'} = -\frac{\partial f'^*}{\partial \alpha_r'} = -\frac{\partial f'}{\partial \alpha_r'} + \frac{\partial \phi}{\partial x_r'}$$

Hamilton-Jacobi equation

$$\Omega\left(-\frac{\partial V}{\partial x}, x\right) = 0 \qquad \Omega^*\left(-\frac{\partial V^*}{\partial x}, x\right) = \Omega\left(-\frac{\partial V^*}{\partial x} - \frac{\partial \phi}{\partial x}, x\right) = 0$$

Slowness 4-vector and slowness equation

$$\sigma_r = -\frac{\partial f}{\partial \alpha_r} \qquad\qquad \sigma_r^* = -\frac{\partial f^*}{\partial \alpha_r} = \sigma_r + \frac{\partial \phi}{\partial x_r}$$

$$\Omega(\sigma, x) = 0 \qquad\qquad \Omega^*(\sigma^*, x) = \Omega\left(\sigma^* - \frac{\partial \phi}{\partial x}, x\right) = 0$$

Here are a few remarks about gauge-invariant properties.

If we start with a given initial 3-wave Σ_0 with equation $F(x) = 0$, then, as remarked above, the directions of the rays are quite different according as we use f or f^*, and there appear to be no interesting gauge-invariant properties in this case. But if we start with a source-event, then the characteristic functions related by the gauge transformation satisfy

$$V^*(x', x) = V(x', x) + \phi(x') - \phi(x), \qquad (2.3.4)$$

in which x_r' are fixed quantities. Suppose that two rays, R_1 and R_2, starting from P' in different directions, meet at an event $P(x)$. The characteristic function is then two-valued, depending on the path, and we have (using suffixes 1 and 2 to distinguish the paths)

$$\left.\begin{array}{l} V_1^*(x', x) = V_1(x', x) + \phi(x') - \phi(x), \\ V_2^*(x', x) = V_2(x', x) + \phi(x') - \phi(x). \end{array}\right\} \qquad (2.3.5)$$

Thus $\qquad V_1^*(x', x) - V_2^*(x', x) = V_1(x', x) - V_2(x', x),$ $\qquad$ (2.3.6)

and so *the action-difference is gauge-invariant.* This is important in connexion with the interference of waves, as discussed by the method of primitive quantization described in Chapter IV.

For the slowness 4-vectors at P we have

$$\sigma^*_{r(1)} = \sigma_{r(1)} + \frac{\partial \phi}{\partial x_r}, \quad \sigma^*_{r(2)} = \sigma_{r(2)} + \frac{\partial \phi}{\partial x_r}. \qquad (2.3.7)$$

Thus, though the slowness 4-vectors are not separately gauge-invariant, their difference is, because we have

$$\sigma^*_{r(1)} - \sigma^*_{r(2)} = \sigma_{r(1)} - \sigma_{r(2)}. \qquad (2.3.8)$$

Singularities on 3-waves are gauge-invariant; for such singularities (corresponding to foci in optics) occur when two rays intersect to the first order, and rays are gauge-invariant.

If our only purpose is to carry over into space-time Hamilton's theory based on a given medium-function $f(x, \alpha)$, then we need not bother about gauge transformations. But when we come to apply the theory to a charged particle in an electromagnetic field, gauge transformations force themselves on our attention, because the appropriate medium-function contains the electromagnetic 4-potential, and that is not completely defined by the field. Now a gauge transformation (in the ordinary sense) of the 4-potential induces in the medium-function a gauge transformation in the sense of (2.3.1), and so, though the rays are definite (gauge-invariant), the waves associated with them are not. Therefore we can get no definite theory of de Broglie waves for a particle in an electromagnetic field until we have decided how to normalize the 4-potential.

It seems best, however, not to mix the Hamiltonian theory based on an assigned $f(x, \alpha)$, which has the inevitability of simplicity, with the question of normalization on which varying opinions may be entertained. Thus though there is a standard way of normalizing the 4-potential (making its divergence vanish), Dirac (1951) has recently proposed another. In so far as de Broglie waves correspond to physical reality, the present theory may be used to test the physical validity of any plan of normalization proposed.

With that, let us drop the question of gauge transformations and proceed on the assumption that the medium-function $f(x, \alpha)$ is given. In the particular applications to electromagnetic fields, the usual simple 4-potentials will be used.

2.4. Geometrical and kinematical descriptions of the unit 3-wave

In Hamilton's optics the structure of all systems of rays possible in a given medium depends solely on the medium-function $v(x, \alpha)$, x denoting position and α the direction cosines of a ray in space. This function is in fact proportional to the refractive index, but depends in general on direction as well as on position. The unit wave in Hamilton's theory is a surface associated with each point, with equation $v(x, y) = 1$ in a notation analogous to that used in (2.1.19). Its interpretation is very simple: the unit wave is the locus after unit time of a light-pulse emitted from the point x, travelling not in the actual medium but in a fictitious homogeneous medium with medium-function $v(x, \alpha)$ *with x held fixed*. This fictitious homogeneous medium (in general anisotropic) is, so to speak, tangent to the real medium at the source x. In a homogeneous medium the distinction between the real medium and the fictitious one disappears. For an isotropic medium such as air the unit waves are spheres, and for a crystal they are Fresnel wave-surfaces.

We have now to carry over this interpretation into space-time, where the medium-function $f(x, \alpha)$ determines the structure of all possible systems of rays. On account of the four-dimensionality of space-time, it is not as easy to do this as one might suppose, and we shall consider two interpretations, the first in terms of Minkowskian geometry and the second by means of a surface moving in the space of a Galilean observer.

Let $P(x)$ be the event with respect to which the unit 3-wave S is taken, S having the equation (2.1.19) in which y_r are coordinates relative to P. Let Q be any event on S (fig. 2.5), so that $y_r = PQ\alpha_r$, PQ being Minkowskian distance and α_r the unit 4-vector along PQ. Then (2.1.19) gives

$$1 = f(x, y) = f(x, PQ\alpha) = PQf(x, \alpha), \qquad (2.4.1)$$

on account of the homogeneity of f, and so

$$PQ = \frac{1}{f(x, \alpha)}. \qquad (2.4.2)$$

This gives us a very simple construction in space-time for the unit 3-wave S: *draw time-like vectors α_r from P into the future and proceed along each through a Minkowskian distance $1/f(x, \alpha)$; the set of events so determined forms the unit 3-wave S at P.*

Thus, for example, if $f(x, \alpha) = \phi(x)(-\alpha_r\alpha_r)^{\frac{1}{2}}$, we get the unit 3-wave at the event x_r by measuring off a Minkowskian distance $1/\phi(x)$ along all the straight lines in space-time drawn into the future from x_r, the unit 3-wave being thus a pseudosphere with that radius. Note that we use straight lines, not the actual rays, which are curved; this corresponds to the use of the fictitious homogeneous medium in Hamilton's optics.

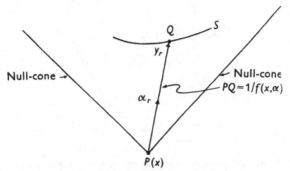

Fig. 2.5. Space-time construction for the unit 3-wave S at the event P.

Let us now describe the unit 3-wave kinematically, using at first Fig. 2.5 as a guide. Let us write its equation more explicitly as

$$f(x; y_1, y_2, y_3, y_4) = 1. \tag{2.4.3}$$

Here y_1, y_2, y_3 are the spatial coordinates of the event Q relative to the event P, and

$$y_4 = ic\tau, \tag{2.4.4}$$

where t is the time at P and $t + \tau$ the time at Q; thus we may also write (2.4.3) as

$$f(x; y_1, y_2, y_3, ic\tau) = 1. \tag{2.4.5}$$

Here x belongs to the fixed event P; y_1, y_2, y_3 and τ vary as we range over the unit 3-wave, satisfying the equation (2.4.5). This equation is naturally interpreted as the history of a *unit 2-wave*, a surface moving in the observer's space, the instantaneous positions of which are obtained by putting $\tau = $ const. *The history of this unit 2-wave is the kinematical description of the unit 3-wave.*

For example, if $f(x, \alpha) = \phi(x)(-\alpha_r \alpha_r)^{\frac{1}{2}}$, then (2.4.5) reads

$$\phi(x)(c^2 \tau^2 - r^2)^{\frac{1}{2}} = 1,$$

where $r^2 = y_1^2 + y_2^2 + y_3^2$. We recognize the unit 2-wave as an expanding sphere, the radius r increasing according to

$$r^2 = c^2 \tau^2 - \phi^{-2},$$

ϕ being a constant depending on the event P with respect to which the unit wave is taken.

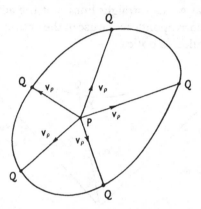

Fig. 2.6. Kinematical description of the unit 3-wave at P as a moving surface of flashes: the lamp of the particle travelling with velocity v_ρ flashes at Q at time τ given by the equation (2.4.10).

There is, however, another kinematical description of the unit 3-wave. A unit vector α_r may be regarded as the velocity 4-vector of a particle moving with velocity v_ρ, where

$$\alpha_\rho = \gamma v_\rho / c, \quad \alpha_4 = i\gamma, \quad \gamma = (1 - v^2/c^2)^{-\frac{1}{2}}, \quad v^2 = v_\rho v_\rho. \quad (2.4.6)$$

We recall that Greek suffixes have the range 1, 2, 3, with the summation convention. Thus the construction of fig. 2.5 may be interpreted kinematically by thinking of particles shot out from a point P in space (fig. 2.6), all at the same time t, in all directions and with all speeds less than c. Each particle travels with constant velocity. The history of any one of these particles is represented by a straight line in space-time, and on this line just one event Q belongs to the unit 3-wave. We may imagine (to make the picture striking) that each particle carries a lamp which is dark all the time except for a flash at the event Q. The 3-wave, or history of the moving 2-wave, is thus represented as a moving surface of flashes. It is of fundamental importance that the velocity of propagation of this surface of flashes is *not* the same as the velocity of the particle carrying the lamp. (The former is the phase velocity and the latter the particle or group velocity in the de Broglie theory.)

To complete this kinematical description of the unit 3-wave in terms of flashing lamps, we have from (2.4.2)

$$PQf(x; \alpha_1, \alpha_2, \alpha_3, \alpha_4) = 1, \qquad (2.4.7)$$

or, by (2.4.6) and the homogeneity of f,

$$\gamma(c^2\tau^2 - r^2)^{\frac{1}{2}} f(x; v_1, v_2, v_3, ic) = c, \qquad (2.4.8)$$

where

$$r^2 = y_\rho y_\rho = v^2\tau^2. \qquad (2.4.9)$$

The equation (2.4.8) may be simplified to

$$\tau f(x; v_1, v_2, v_3, ic) = 1. \qquad (2.4.10)$$

This equation describes the moving unit 2-wave in the sense that it gives the flashing time τ for the lamp carried by the particle which has the velocity v_ρ.

Returning to the example $f(x, \alpha) = \phi(x)(-\alpha_r \alpha_r)^{\frac{1}{2}}$, we note that for this case (2.4.10) reads

$$\tau\phi(x)(c^2 - v^2)^{\frac{1}{2}} = 1. \qquad (2.4.11)$$

We observe that if the particle has zero velocity, its lamp flashes at $\tau = 1/c\phi$, and that for a particle with velocity close to c the time to flashing is very long.

2.5. Wave velocity and ray velocity

Consider a 3-wave Σ_3 with space-time equation

$$F(x) = 0. \qquad (2.5.1)$$

This is not the (artificial) unit 3-wave we have been discussing. It is any 3-space, subject only to the general restriction mentioned in §2.2.

The above equation represents the history of a moving 2-wave Σ_2, the instantaneous positions of which are found by putting $x_4 = $ const. We have then two pictures: a static picture (fig. 2.7) of Σ_3 in space-time, and a moving picture (fig. 2.8) of Σ_2 in the observer's space. There is not much to be said about the former except to remember that the slowness 4-vector σ_r is normal to Σ_3. It is the latter that requires discussion.

In fig. 2.8 let Σ_2 be the position of the 2-wave at time t and Σ_2' its position at time $t + dt$. Let n_ρ be the unit normal to Σ_2, pointing in the sense of propagation, and let u_ρ be the *wave velocity* 3-*vector*, defined by the equation

$$u_\rho\, dt = ABn_\rho, \qquad (2.5.2)$$

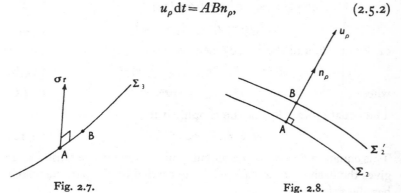

Fig. 2.7.   Fig. 2.8.

Fig. 2.7. Space-time diagram of 3-wave Σ_3 and the slowness 4-vector σ_r normal to it: the events A and B appear also in fig. 2.8.

Fig. 2.8. Kinematical picture of the 2-wave Σ_2 advancing in space with velocity u_ρ.

where AB is the length of the normal displacement from Σ_2 to Σ_2', this infinitesimal vector being

$$dx_\rho = ABn_\rho. \qquad (2.5.3)$$

We seek to calculate u_ρ in terms of the partial derivatives of the function F. Since n_ρ are proportional to $F_{,\rho}$ (we shall use commas to indicate partial derivatives), we have

$$u_\rho = \phi F_{,\rho}, \qquad (2.5.4)$$

where ϕ is a factor of proportionality. But the space-time displacement from the event at A to the event at B lies in Σ_3, and so

$$F_{,\rho}\, dx_\rho + F_{,4}\, dx_4 = 0. \qquad (2.5.5)$$

Division by dt gives
$$F_{,\rho} u_\rho + icF_{,4} = 0, \qquad (2.5.6)$$

and hence, on substitution from (2.5.4),

$$\phi = -ic\,\frac{F_{,4}}{F_{,\pi}F_{,\pi}}. \qquad (2.5.7)$$

Putting this value into (2.5.4), we get the following expression for the velocity of the 3-wave $F(x) = 0$:

$$u_\rho = -ic\frac{F_{,\rho}F_{,4}}{F_{,\pi}F_{,\pi}}. \qquad (2.5.8)$$

For example, for a plane wave proceeding along the x_1-axis, we have

$$F(x) = a_1 x_1 + a_4 x_4 + a = 0,$$

the a's being constants; we verify by (2.5.8)

$$u_1 = -ica_4/a_1, \quad u_2 = 0, \quad u_3 = 0.$$

For a spherical 2-wave expanding uniformly, we have

$$2F(x) = a^2 x_\rho x_\rho + b^2 x_4^2 = 0.$$

Then $\quad F_{,\rho} = a^2 x_\rho, \quad F_{,4} = b^2 x_4, \quad F_{,\pi}F_{,\pi} = a^4 x_\pi x_\pi = -a^2 b^2 x_4^2,$

and (2.5.8) gives

$$u_\rho = -ic\frac{b^2 x_\rho x_4}{a^2 x_\pi x_\pi} = ic\frac{x_\rho}{x_4} = \frac{cb}{a}\frac{x_\rho}{(x_\pi x_\pi)^{\frac{1}{2}}}.$$

This example serves to remind us that in the case of a single 3-wave $F(x) = 0$ there is of course no unique formal expression for u_ρ in terms of the space-time coordinates, because u_ρ exists only on the 3-wave $F(x) = 0$.

To find the slowness 4-vector corresponding to the 3-wave (2.5.1), we have, since it is normal to the 3-wave,

$$\sigma_r = \theta F_{,r}, \qquad (2.5.9)$$

where θ is a factor of proportionality, determined by the slowness equation

$$\Omega(\sigma, x) = \Omega\left(\theta\frac{\partial F}{\partial x}, x\right) = 0. \qquad (2.5.10)$$

In view of (2.5.8) and (2.5.9), an observer can calculate the wave velocity in terms of the slowness 4-vector by the formula

$$u_\rho = -ic\frac{\sigma_\rho \sigma_4}{\sigma_\pi \sigma_\pi}; \qquad (2.5.11)$$

the square is

$$u^2 = -c^2\frac{\sigma_4^2}{\sigma_\pi \sigma_\pi}. \qquad (2.5.12)$$

As an example, consider $f(x, \alpha) = \phi(x)(-\alpha_r \alpha_r)^{\frac{1}{2}}$, which gives $\Omega(\sigma, x) = \sigma_r \sigma_r + \phi^2 = 0$, and so $\sigma_\pi \sigma_\pi = \phi^2 - \sigma_4^2$. Then (2.5.12) gives

$$u^2 = c^2\frac{\sigma_4^2}{\sigma_4^2 - \phi^2}.$$

This example illustrates a fact to be discussed below. A knowledge of $f(x, \alpha)$ does not suffice to find the speed of wave propagation; in this particular example f determines u to within one adjustable parameter, σ_4. To find that parameter, we must be told about the 3-wave (2.5.1) in order to use (2.5.9).

Suppose now that we no longer think of some given 3-wave, but rather ask the sort of question we might ask in optics: what is the wave velocity for a wave propagated along a given line L in space? To answer this question, we specify a position on L and a time (that is, we assign x_r), and we denote by n_ρ a unit vector pointing along L in the sense in which we wish the propagation to take place. We wish to see a moving 2-wave with normal n_ρ, i.e. with u_ρ pointing in that direction, and we note from (2.5.11) that the appropriate slowness 4-vector must satisfy

$$\sigma_\rho = \psi n_\rho, \qquad (2.5.13)$$

where ψ is a factor of proportionality. Now we have the slowness equation $\Omega(\sigma, x) = 0$, or more explicitly

$$\Omega(\sigma_1, \sigma_2, \sigma_3, \sigma_4; x) = 0, \qquad (2.5.14)$$

and when we substitute in this from (2.5.13) we get

$$\Omega(\psi n_1, \psi n_2, \psi n_3, \sigma_4; x) = 0. \qquad (2.5.15)$$

This is *one* equation for *two* unknowns, ψ and σ_4, and in the resulting indeterminacy lies an essential property of de Broglie waves. Space is a dispersive medium, in the sense that there is not a unique wave velocity in a given direction, but a range of wave velocities corresponding to arbitrary choice of σ_4. This is the fourth component of a 4-vector and so is the analogue of frequency or energy, in the sense that it transforms in the same way under a Lorentz transformation, but the concepts of frequency and energy have not so far appeared in this mathematical theory.

Recognizing then that σ_4 is arbitrary, we answer the question raised above by using (2.5.11) and (2.5.13); the wave velocity along L is the 3-vector

$$u_\rho = -icn_\rho \sigma_4 / \psi, \qquad (2.5.16)$$

where ψ is to be found to satisfy (2.5.15), n_ρ being the direction cosines of L.

Connexions between wave and ray velocities. Returning to (2.5.11), we note that it can be used to give the wave velocity u_ρ in terms of the ray velocity v_ρ, the latter being the velocity of a particle having

THEORY OF RAYS AND WAVES IN SPACE-TIME 29

a ray for world line, and so being given in terms of α_r by (2.4.6). For $\sigma_r = -\partial f/\partial\alpha_r$, a known function of x_r, α_r and hence a known function of x_r, v_ρ. The relation between u_ρ and v_ρ is to be regarded as a generalized form of the de Broglie relationship between phase velocity and particle velocity, for if we think of the present mathematical theory in physical terms, the waves are phase waves and the rays possible histories of particles.

To pursue the connexion further, by (2.1.33) we have

$$\frac{\alpha_\rho}{\alpha_4} = \frac{\partial\Omega/\partial\sigma_\rho}{\partial\Omega/\partial\sigma_4}, \qquad (2.5.17)$$

and so by (2.4.6) the ray velocity is expressed in terms of the slowness 4-vector by

$$v_\rho = ic\frac{\partial\Omega/\partial\sigma_\rho}{\partial\Omega/\partial\sigma_4}. \qquad (2.5.18)$$

The whole question of the connexion between the wave velocity u_ρ and the ray velocity v_ρ is thus contained in the seven equations

$$u_\rho = -ic\frac{\sigma_\rho\sigma_4}{\sigma_\pi\sigma_\pi}, \quad v_\rho = ic\frac{\partial\Omega/\partial\sigma_\rho}{\partial\Omega/\partial\sigma_4}, \quad \Omega(\sigma,x)=0. \quad (2.5.19)$$

By elimination of σ_r we get three equations connecting u_ρ and v_ρ (and of course x_r), and these may be solved so as to express u_ρ in terms of v_ρ, or vice versa. But of course we cannot carry out these algebraic processes explicitly in the general case; we shall do so in Chapter III for the case of a particle which is free or moves in a given electromagnetic field.

There is another way of getting the connexion between the two velocities. By virtue of (2.1.9) and (2.1.33) we may also write (2.5.19) in the form

$$u_\rho = -ic\frac{\dfrac{\partial f}{\partial\alpha_\rho}\dfrac{\partial f}{\partial\alpha_4}}{\dfrac{\partial f}{\partial\alpha_\pi}\dfrac{\partial f}{\partial\alpha_\pi}}, \quad v_\rho = ic\frac{\alpha_\rho}{\alpha_4}. \qquad (2.5.20)$$

Only the ratios of the α's appear here, and the elimination of these three ratios from the six equations (2.5.20) leaves us with three equations connecting u_ρ and v_ρ.

As an example consider $f(x,\alpha) = \phi(x)(-\alpha_r\alpha_r)^{\frac{1}{2}}$, giving

$$\Omega(\sigma,x) = \sigma_r\sigma_r + \phi^2 = 0.$$

Then (2.5.19) gives

$$u_\rho = -ic\frac{\sigma_\rho \sigma_4}{\sigma_\pi \sigma_\pi}, \quad v_\rho = ic\frac{\sigma_\rho}{\sigma_4}.$$

Hence $ic\sigma_\rho = v_\rho \sigma_4$, and so $\Omega = 0$ gives

$$v^2\sigma_4^2 - c^2\sigma_4^2 - c^2\phi^2 = 0, \quad \sigma_4 = i\gamma\phi, \quad \gamma = (1 - v^2/c^2)^{-\frac{1}{2}};$$

therefore $\sigma_\rho = v_\rho \gamma\phi/c$ and hence

$$u_\rho = -ic\frac{v_\rho \gamma\phi}{c}\frac{i\gamma\phi c^2}{v^2\gamma^2\phi^2} = \frac{c^2 v_\rho}{v^2},$$

so that $u^2 = c^4/v^2$ or $uv = c^2$, the well-known de Broglie relationship between the two velocities. It is an accidental feature of the example considered that u_ρ and v_ρ have the same direction in space; in general they do not. The result comes out as simply by applying (2.5.20), which read

$$u_\rho = -ic\frac{\alpha_\rho \alpha_4}{\alpha_\pi \alpha_\pi}, \quad v_\rho = ic\frac{\alpha_\rho}{\alpha_4}.$$

2.6. Wave packets and group velocity

The concepts of wave packet and group velocity appear in current physical thought to be bound up with the interference of waves and Fourier transforms. In the present discussion we go back to much more primitive ideas.

Consider the rays and waves from a source-event, or more generally the rays and waves from a given initial 3-wave Σ_0 (§ 2.2). Let us pick out a small *bundle of rays* (as we so often do in optics) by marking off on Σ_0 a small three-dimensional region and omitting all the rays except those that come from this small region. The ray-bundle fills a thin tube in space-time, and the portions of the 3-waves contained in this tube we call a *wave packet*. Fig. 2.9 shows a portion of the tube and the 3-waves inside it.

Let us now look at this kinematically. Each ray appears to a Galilean observer as a moving point, and so the ray-bundle appears as a small cloud of points, all moving with approximately the same velocity v_ρ. Let us denote by $R(t)$ the small region in space occupied by this cloud at time t; it is given by cutting the space-time ray-bundle by the hyperplane $t = \text{const.}$, as shown in fig. 2.9 for $t = t_1$ and $t = t_2$.

At time t there are 2-waves in $R(t)$ but none outside it, for we have cut off the 3-waves at the boundary of the ray-bundle. The 2-waves are sections of the 3-waves by $t = $ const., and they move with the velocity u_ρ given by (2.5.11). This is of course different from the

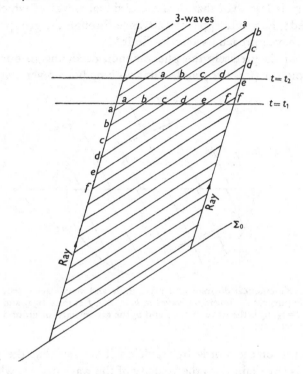

Fig. 2.9. Space-time diagram of a ray-bundle and the associated wave packet, including the 3-waves a, b, c, d, e, f.

ray velocity v_ρ. On the boundary of $R(t)$ (a moving surface in the observer's space) 2-waves are constantly disappearing and new ones appearing.

This is illustrated in fig. 2.10, in which for graphical purposes the region $R(t)$ is cut into two parts by a plane containing the ray velocity v_p, one half being pictured at time t_1 and the other at time t_2. Each half is filled with a continuum of 2-waves from which we pick out a few at time t_1 in order to follow their motion. They are labelled a, b, c, d, e, f, these labels belonging to certain 3-waves as shown in fig. 2.9. At time t_2 the 2-waves e and f have disappeared and two new waves have appeared.

Lord Rayleigh (1877) once wrote: 'It has often been remarked that, when a group of waves advances into still water, the velocity of the group is less than that of the individual waves of which it is composed; the waves appear to advance through the group, dying away as they approach its anterior limit.'

Now this is precisely the phenomenon occurring in our kinematical picture (fig. 2.10), and so it is nothing but a reversion to the

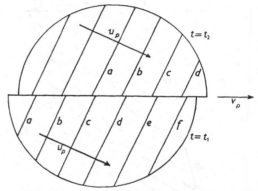

Fig. 2.10. Kinematical diagram of a ray-bundle and wave packet (cut in two for graphic purposes), showing 2-waves a, b, c, d, e, f at time $t = t_1$ and a, b, c, d at time $t = t_2$: u_p is the wave velocity and v_p the ray velocity, or group velocity, or particle velocity.

original use of the words by Rayleigh if we say that the group velocity is (by definition) the velocity of the wave packet, which is itself the ray velocity. If we read *particle velocity* for *ray velocity*, then the de Broglie relationship

group velocity of waves = particle velocity

becomes a mere truism in the present theory.

But it is interesting to see how the familiar formula for group velocity fits into the theory. This formula reads

$$\frac{1}{w} = \frac{\partial}{\partial v}\left(\frac{v}{u}\right), \qquad (2.6.1)$$

where w is the group velocity, u the wave velocity and v the frequency, u being regarded as a function of v in a dispersive medium.

We turn to (2.5.16) which gives the wave velocity in the direction n_ρ in terms of σ_4, and obtain the magnitude u of this 3-vector as

$$u = -\mathrm{i}c\sigma_4/\psi, \quad \sigma_4/u = \mathrm{i}\psi/c. \tag{2.6.2}$$

Now n_ρ being given, (2.5.15) defines ψ as a function of σ_4, and so differentiation of (2.6.2) gives

$$\frac{\partial}{\partial \sigma_4}\left(\frac{\sigma_4}{u}\right) = \frac{\mathrm{i}}{c}\psi'(\sigma_4), \tag{2.6.3}$$

where, by (2.5.15), $\dfrac{\partial \Omega}{\partial \sigma_\rho} n_\rho \psi'(\sigma_4) + \dfrac{\partial \Omega}{\partial \sigma_4} = 0.$ (2.6.4)

But, by (2.1.33), $\partial\Omega/\partial\sigma_r$ are proportional to α_r, and if we make use of (2.4.6) we get from (2.6.4)

$$v_\rho n_\rho \psi'(\sigma_4) + \mathrm{i}c = 0, \quad \psi'(\sigma_4) = -\frac{\mathrm{i}c}{v_\rho n_\rho}. \tag{2.6.5}$$

Substitution in (2.6.3) then gives

$$\frac{1}{v_\rho n_\rho} = \frac{\partial}{\partial \sigma_4}\left(\frac{\sigma_4}{u}\right). \tag{2.6.6}$$

Comparing this with (2.6.1), we see that *the component $v_\rho n_\rho$ of ray velocity in the direction of propagation of a 2-wave (i.e. normal to it) is given by the usual formula for group velocity, provided that σ_4 is interpreted as being proportional to frequency.*

The appearance of the formula (2.6.6), formally the same as (2.6.1), in a theory in which there is no *interference* (because there is no phase in it—that will not come until Chapter IV) makes one wonder whether the whole question of group velocity has not been unduly complicated in the usual treatment by thinking in terms of kinematics in space rather than in terms of the geometry of space-time.

To illustrate the meaning of (2.6.6), let us again take as an example $f(x, \alpha) = \phi(x)(-\alpha_r\alpha_r)^{\frac{1}{2}}$, $\Omega(\sigma, x) = \sigma_r\sigma_r + \phi^2 = 0$. Now (2.5.15) gives $\psi^2 + \sigma_4^2 + \phi^2 = 0$, $\psi = (-\sigma_4^2 - \phi^2)^{\frac{1}{2}}$,

and so by (2.5.16) $\quad u^2 = -\dfrac{c^2\sigma_4^2}{\psi^2} = \dfrac{c^2\sigma_4^2}{\sigma_4^2 + \phi^2}.$

Hence $\quad\dfrac{\sigma_4}{u} = \dfrac{\mathrm{i}}{c}(-\sigma_4^2 - \phi^2)^{\frac{1}{2}},$

and (2.6.6) gives

$$\frac{1}{v_\rho n_\rho} = \frac{\partial}{\partial \sigma_4}\left(\frac{\sigma_4}{u}\right) = \frac{-i\sigma_4}{c(-\sigma_4^2 - \phi^2)^{\frac{1}{2}}} = \frac{u}{c^2}.$$

This checks with the result established for this particular example at the end of §2.5.

2.7. Laws of reflexion and refraction

The variational principle (2.1.3) provides us not only with the Euler-Lagrange equations (2.1.4) for a ray, but also with the abrupt change occurring when a ray meets a 3-space across which the

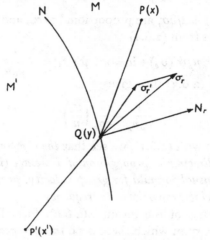

Fig. 2.11. Refraction at a 3-space of discontinuity N: the 4-vector $\sigma_r - \sigma_r'$ is parallel to the normal to N at Q.

medium-function $f(x, \alpha)$ is discontinuous. Such changes correspond to *reflexion* and *refraction*, reflexion occurring if the ray turns back and does not cross the 3-space of discontinuity, and refraction if it does cross it.

To discuss these phenomena, let N be the 3-space of discontinuity, with equation $N(x) = 0$. Let us put

$$N_r = \frac{\partial N}{\partial x_r}, \qquad (2.7.1)$$

so that N_r is a 4-vector normal to N (fig. 2.11). N divides spacetime into two regions, M' (the initial region) and M (the final region), with medium-functions $f'(x', \alpha')$ and $f(x, \alpha)$ respectively.

Consider a ray proceeding from $P'(x')$ in M' to $P(x)$ in M, crossing N at $Q(y)$. Now by (2.1.3) we have

$$\delta \int_{P'}^{P} f \, ds = 0, \qquad (2.7.2)$$

P' and P being held fixed. Let us apply a particular variation in which Q is shifted on N from y_r to $y_r + \delta y_r$, the two parts of the varied curve being rays in M' and in M respectively. Then (2.7.2) gives

$$\delta \int_{P'}^{Q} f'(x', \alpha') \, ds' + \delta \int_{Q}^{P} f(x, \alpha) \, ds = 0, \qquad (2.7.3)$$

and so by (2.1.6), since Q is the final event for the first integral and the initial event for the second,

$$\frac{\partial f'}{\partial \alpha'_r} \delta y_r - \frac{\partial f}{\partial \alpha_r} \delta y_r = 0,$$

or
$$(\sigma_r - \sigma'_r) \delta y_r = 0. \qquad (2.7.4)$$

This holds for all variations δy_r in N, i.e. for all δy_r satisfying

$$N_r \delta y_r = 0, \qquad (2.7.5)$$

and so we have
$$\sigma_r - \sigma'_r = k N_r, \qquad (2.7.6)$$

where k is a factor of proportionality. *This equation* (2.7.6) *is the general law of reflexion and refraction.*

Let us see how this law is to be used to obtain the reflected or refracted ray. We regard the incident ray and the event of incidence as given, and so α'_r and y_r are given. From these values we obtain the incident slowness 4-vector σ'_r by means of $\sigma'_r = -\partial f'/\partial \alpha'_r$. N_r is given by (2.7.1), evaluated at y_r. Then in (2.7.6) and the equation

$$\Omega(\sigma, y) = 0, \qquad (2.7.7)$$

characterizing the final medium M, we have five equations for the five unknowns σ_r, k. When σ_r has been found, we get the final ray by (2.1.33):

$$\alpha_r = \theta \frac{\partial \Omega}{\partial \sigma_r}, \quad \alpha_r \alpha = -1, \qquad (2.7.8)$$

these being five equations for α_r, θ.

That is the programme for refraction. For reflexion, we must use $\Omega'(\sigma, y) = 0$, characterizing the initial medium, since the rays go back into it, instead of (2.7.7).

A complete discussion of reflexion and refraction is facilitated by using space-time diagrams. The geometrical meaning of the law (2.7.6) is very simple. It tells that *if we draw from Q the 4-vectors σ_r and σ'_r, the line joining their extremities is parallel to the normal to N at Q, i.e. parallel to N_r.* We may also say that the tangential components of these 4-vectors are unchanged by reflexion or refraction.

To find out whether reflexion or refraction (or both or neither) occur, we make a space-time diagram as in fig. 2.12, showing the event of incidence Q and the normal N_r, and also the 3-surfaces of slowness relative to Q, these being as in (2.1.20),

$$W' \text{ for } M' : \Omega'(\sigma', y) = 0; \quad W \text{ for } M : \Omega(\sigma, y) = 0. \quad (2.7.9)$$

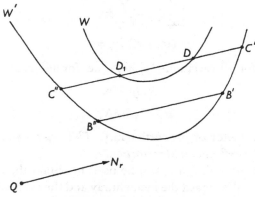

Fig. 2.12. The 3-surfaces of slowness for reflexion and refraction, W' being that for the initial region M' and W that for the final region M.

We draw from Q the incident slowness 4-vector σ'_r. Its extremity must lie on W'; let it be at B'. To get the final slowness 4-vector σ_r, we are to draw through B' a line parallel to N_r and, in the case of refraction, seek its intersection with W.

It is not feasible to deal with all the possibilities depending on the forms of W' and W. Fig. 2.12 will serve as an illustration. In the case shown there is no such intersection, and so refraction is impossible.

But the line in question does cut W' at B'', and so QB'' gives us a final slowness 4-vector suited to a ray in M'. Thus we have reflexion, and in fact total reflexion, since refraction has been ruled out. But before we can be sure that this reflexion is valid, we must make sure of two things:

(i) the reflected ray must carry us back into M' and not on into M;

(ii) the reflected ray must point into the future and not into the past.

If the ray obtained from QB'' should violate either of these conditions, then all we can say is that both reflexion and refraction are impossible.

Consider now the case where the incident slowness 4-vector is QC' in fig. 2.12. Now the line drawn through C' parallel to N_r cuts W at D and D_1 and it also cuts W' at C''. We have now two possibilities for refracted rays, corresponding to the slowness 4-vectors QD and QD_1, and one possibility for a reflected ray, corresponding to the slowness QC''. The reflected ray must be examined relative to the conditions (i) and (ii) above, and the refracted rays for similar conditions, viz.

(i) the refracted ray must carry us on into M and not back into M';

(ii) the refracted ray must point into the future and not into the past.

It will be clear from the above discussion that the general law (2.7.6) may give us invalid or extraneous solutions; these must be eliminated by examining each problem separately.

Types of 3-spaces of discontinuity. The normal N_r of the 3-space N may be time-like, space-like or null.

A time-like N_r means that N is the history of a 2-space moving faster than light. For the speed of propagation u is, by (2.5.8), such that

$$u^2 = -c^2 \frac{N_4^2}{N_\pi N_\pi}, \tag{2.7.10}$$

and so

$$\frac{u^2}{c^2} - 1 = -\frac{N_r N_r}{N_\pi N_\pi} > 0, \tag{2.7.11}$$

provided N_r is time-like.

This does not violate the relativistic limitation of velocities to lie below c, because the velocity in question is not the velocity of a particle or of a signal. We can give an optical analogy to illustrate a time-like N_r: suppose light is passing through a lake of water, and that the water freezes suddenly throughout its whole volume. The history of the water before freezing gives a region M' in space-time and after freezing a region M, the optical properties changing suddenly across a 3-space N with equation $t = \text{const}$. It is not so easy to think of a case of time-like N_r applicable to geometrical mechanics, but it is well to keep the possibility of a time-like N_r in the general theory.

A space-like N_r occurs when N is the history of a 2-space (surface) moving with a speed less than c. This is a natural case to consider, particularly if the 2-space is fixed in the space of some Galilean observer. A potential-jump, to be considered later, provides an example.

A null N_r means that N is the history of a 2-space travelling with the speed of light. This would occur if a particle passed into an electromagnetic shock wave.

Example. As an illustrative example, consider regions specified by

$$M': \quad f'(x', \alpha') = \phi'(x')(-\alpha'_r \alpha'_r)^{\frac{1}{2}},$$
$$\Omega'(\sigma', x') = \sigma'_r \sigma'_r + \phi'^2 = 0;$$
$$M: \quad f(x, \alpha) = \phi(x)(-\alpha_r \alpha_r)^{\frac{1}{2}},$$
$$\Omega(\sigma, x) = \sigma_r \sigma_r + \phi^2 = 0.$$

Let us first take for N the hyperplane $x_4 = 0$, so that $N_\rho = 0$, $N_4 = 1$. We have $x'_4/i < 0$ in M' and $x_4/i > 0$ in M. Then (2.7.6) gives $\sigma_\rho = \sigma'_\rho$, and σ_4 is found from

$$\sigma_4^2 = -\sigma'_\rho \sigma'_\rho - \phi^2.$$

We have $\alpha_r = \sigma_r/\phi$, and so the refracted ray is

$$\alpha_\rho = \sigma'_\rho/\phi, \quad \alpha_4 = i(\sigma'_\rho \sigma'_\rho + \phi^2)^{\frac{1}{2}}/\phi.$$

This satisfies the requirements (i) and (ii) above. If we seek a *reflected* slowness 4-vector σ_r, we get $\sigma_\rho = \sigma'_\rho$, but this is not admissible because we get a ray either pointing on into M or pointing into the past. Indeed, it is clear that we can never have reflexion from a 3-space of discontinuity if N_r is time-like.

Let us now change N, taking for N_r a space-like 4-vector. We can then make $N_4 = 0$ by choice of frame of reference. Then (2.7.6) gives

$$\sigma_\rho - \sigma'_\rho = kN_\rho, \quad \sigma_4 = \sigma'_4.$$

Hence $\Omega = 0$ gives

$$(\sigma'_\rho + kN_\rho)(\sigma'_\rho + kN_\rho) + \sigma'^2_4 + \phi^2 = 0,$$

and we get for k the quadratic equation

$$k^2 N_\rho N_\rho + 2kN_\rho \sigma'_\rho + \phi^2 - \phi'^2 = 0.$$

If this has real roots, we get a unique refracted ray, the second solution being extraneous (it gives a ray passing back into M'). If the roots are imaginary, there is no refraction. For reflexion we have

$$\sigma_\rho - \sigma'_\rho = kN_\rho, \quad \sigma_4 = \sigma'_4,$$

and $\Omega'\,(\sigma, y) = 0$ gives

$$(\sigma'_\rho + kN_\rho)(\sigma'_\rho + kN_\rho) + \sigma'^2_4 + \phi'^2 = 0,$$

or, since $\sigma'_\rho \sigma'_\rho + \sigma'^2_4 + \phi'^2 = 0$,

$$2kN_\rho \sigma'_\rho + k^2 N_\rho N_\rho = 0.$$

Rejecting $k = 0$ as extraneous for reflexion, we get a reflected ray by taking

$$k = -2 \frac{N_\rho \sigma'_\rho}{N_\pi N_\pi},$$

and the unit 4-vector along this ray is

$$\alpha_\rho = \frac{\sigma'_\rho + kN_\rho}{\phi'} = \alpha'_\rho + \frac{kN_\rho}{\phi'}, \quad \alpha_4 = \frac{\sigma_4}{\phi'} = \alpha'_4.$$

Refraction through a hole. In the preceding theory we have taken N to be a 3-space. But there are other cases of interest. Suppose that all space is divided into two parts by a wall. The histories of these two parts give us two space-time regions, M' and M, and there is no communication between them on account of the wall.

But now let a small hole be made in the wall. The history of this hole is a curve in space-time, and this curve establishes a one-dimensional connexion between M' and M. Similarly, if we make a slit in the wall, we get a two-dimensional region through which we can pass from M' to M.

We can explore refraction through a hole or a slit by means of the variational principle (2.1.3). Let us take the case of a hole. Its history may be represented by four equations

$$y_r = y_r(w), \tag{2.7.12}$$

where w is a parameter, and this curve has a tangent 4-vector

$$A_r = \frac{dy_r}{dw}. \tag{2.7.13}$$

Now by the variational principle we are again led to (2.7.4), to be satisfied for variation of y_r on (2.7.12), but instead of (2.7.6) we now get
$$(\sigma_r - \sigma_r') A_r = 0. \tag{2.7.14}$$

In order to find the refracted slowness 4-vector σ_r (and hence the refracted ray direction α_r), we have now only *two* equations, viz. (2.7.14) and
$$\Omega(\sigma, x) = 0. \tag{2.7.15}$$

Thus, *for the passage of a ray through a hole, there is a twofold arbitrariness in the emergent slowness 4-vector, and hence a twofold arbitrariness in the space-time direction of the emergent ray*. In the case of a slit, there is a onefold arbitrariness.

For example, suppose that there is the same medium-function for M' and M: $f(x, \alpha) = (-\alpha_r \alpha_r)^{\frac{1}{2}}$. Suppose that the history of the hole is parallel to the t-axis, i.e. the hole is at rest. Then $A_\rho = 0$ and (2.7.14) gives $\sigma_4 - \sigma_4' = 0$. Hence σ_ρ are arbitrary except for
$$\sigma_\rho \sigma_\rho = -1 - \sigma_4'^2.$$

Thus, since $\alpha_r = \sigma_r$, we have
$$\alpha_\rho \alpha_\rho = -1 - \alpha_4'^2,$$

and so by (2.4.6) the velocity v_ρ of the emergent ray is conditioned only by
$$v_\rho v_\rho \gamma^2/c^2 = -1 + \gamma'^2,$$

which leads at once to $v = v'$. In fact, for this example, rays emerge from the hole in all spatial directions, with a speed equal to that of the incident ray.

2.8. Hamilton's T-function

Hamilton's method was long regarded as useless in practical optics on account of the difficulty of calculating his characteristic function V. His other characteristic function T was overlooked until rediscovered by Bruns, but now (often under the name of 'angle eikonal') provides the most systematic method for the discussion of the aberrations of optical instruments. Let us now introduce the T-function into the relativistic theory of rays.

Let space-time be divided into three parts, M', N and M, as in fig. 2.13, and let rays pass from M' through N into M. Although the T-function exists under quite general conditions, it is most useful in the case of homogeneity, and so we shall suppose that M' and M (but not necessarily N) are *homogeneous* in the sense that in each of them the medium-function $f(x, \alpha)$ is independent of x_r. Then the Euler-Lagrange equations (2.1.4) tell us at once that the rays in M' and in M are straight.

Let $V(x', x)$ be the characteristic function (2.1.5) for a ray which goes from $P'(x')$ in M' to $P(x)$ in M. We define T by writing

$$T = V(x', x) + \sigma_r x_r - \sigma'_r x'_r, \quad (2.8.1)$$

where σ'_r and σ_r are the slowness 4-vectors at P' and P respectively. These vectors satisfy

$$\sigma'_r = -\frac{\partial f'}{\partial \alpha'_r}, \quad \sigma_r = -\frac{\partial f}{\partial \alpha_r}, \quad (2.8.2)$$

and they are therefore constant along the parts of the ray in M' and M,

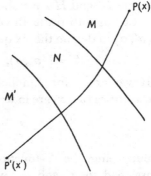

Fig. 2.13. Homogeneous regions M', M of space-time, separated by a region N.

respectively, since α'_r and α_r are constant, the parts being straight.

Let us now vary from the ray $P'P$ to a neighbouring ray with slightly different end-events. Remembering that

$$\frac{\partial V}{\partial x'_r} = \sigma'_r, \quad \frac{\partial V}{\partial x_r} = -\sigma_r, \quad (2.8.3)$$

we get

$$\delta T = x_r \delta \sigma_r - x'_r \delta \sigma'_r. \quad (2.8.4)$$

Thus the value of T is unchanged if we slide P' and P along the ray, for then the slowness 4-vectors are unchanged. In fact, T is a 'function of the ray', not involving actually the choice of the initial and final events on it, and so, since a ray is in general determined by its initial and final slowness 4-vectors, we may write

$$T = T(\sigma'_1, \sigma'_2, \sigma'_3, \sigma'_4, \sigma_1, \sigma_2, \sigma_3, \sigma_4). \quad (2.8.5)$$

But these eight arguments are not independent. Since $f(x, \alpha)$ is independent of x_r in M, so also is the function $\Omega(\sigma, x)$ of (2.1.14), and hence the slowness equations for M' and M read

$$\Omega'(\sigma'_1, \sigma'_2, \sigma'_3, \sigma'_4) = 0, \quad \Omega(\sigma_1, \sigma_2, \sigma_3, \sigma_4) = 0. \quad (2.8.6)$$

It is convenient to solve these equations for σ_4' and σ_4; since these are the fourth components of 4-vectors, they are pure imaginaries, and so we shall write the solutions of (2.8.6) in the form

$$\sigma_4' = iH'(\sigma_1', \sigma_2', \sigma_3'), \quad \sigma_4 = iH(\sigma_1, \sigma_2, \sigma_3), \qquad (2.8.7)$$

where H' and H are real.

Let us substitute these in (2.8.5), using an umbral notation (σ', σ) to denote the six quantities σ_ρ', σ_ρ. Thus we have

$$T = T(\sigma', \sigma), \quad \sigma_4' = iH'(\sigma'), \quad \sigma_4 = iH(\sigma). \qquad (2.8.8)$$

If we omit, for simplicity, certain exceptional cases, the six quantities (σ', σ) are independent, and so (2.8.4) gives

$$\frac{\partial T}{\partial \sigma_\rho'} = -x_\rho' - ix_4' \frac{\partial H'}{\partial \sigma_\rho'}, \quad \frac{\partial T}{\partial \sigma_\rho} = x_\rho + ix_4 \frac{\partial H}{\partial \sigma_\rho}. \qquad (2.8.9)$$

Supposing the T-function known, and constant values being assigned to σ_ρ' and σ_ρ, we have in the first set of (2.8.9) three linear equations for the initial portion of the ray and in the second set three linear equations for the final portion of the ray.

It is important to note that (2.8.9) holds only if the six quantities σ_ρ', σ_ρ are independent. We shall presently discuss cases where this is not so, and give the modified form of (2.8.9).

Example. We shall later consider how $T(\sigma', \sigma)$ is to be calculated. For the moment it may be regarded as an arbitrary given function of its six arguments, and $H'(\sigma')$, $H(\sigma)$ may be regarded also as arbitrary functions. To illustrate the meaning of (2.8.9), let us take

$$T(\sigma', \sigma) = \tfrac{1}{2}(\sigma_\rho' - \sigma_\rho)(\sigma_\rho' - \sigma_\rho),$$

$$\Omega'(\sigma') = \sigma_r' \sigma_r' + k'^2 = 0, \quad \sigma_4' = i(\sigma_\rho' \sigma_\rho' + k'^2)^{\frac{1}{2}} = iH'(\sigma'),$$

$$\Omega(\sigma) = \sigma_r \sigma_r + k^2 = 0, \quad \sigma_4 = i(\sigma_\rho \sigma_\rho + k^2)^{\frac{1}{2}} = iH(\sigma).$$

Then (2.8.9) gives for the rays

$$\sigma_\rho' - \sigma_\rho = -x_\rho' - ix_4' \sigma_\rho'/H', \quad \sigma_\rho - \sigma_\rho' = x_\rho + ix_4 \sigma_\rho/H.$$

Suppose now that we take an initial ray passing through the two events $(x_r' = 0)$ and $(x_\rho' = 0, x_4' = i)$. Then by the first of the above equations, the σ_ρ', σ_ρ for the ray satisfy

$$\sigma_\rho' - \sigma_\rho = 0, \quad \sigma_\rho' - \sigma_\rho = \sigma_\rho'/H'.$$

Thus $\sigma_\rho' = \sigma_\rho = 0$. The second set of (2.8.9) then tells us that the final ray has the equations $x_\rho = 0$.

We see that *any* initial ray through $x'_r = 0$ satisfies $\sigma'_\rho = \sigma_\rho$, and so the equations of the final ray are

$$x_\rho + \mathrm{i}x_4\,\sigma_\rho/H = 0.$$

But, for arbitrary σ_ρ, this represents the totality of all straight rays through $x_r = 0$, and so, for this example, the intermediate region N has the effect of transforming the congruence of initial rays through the space-time origin into a congruence of final rays through this origin. This does not mean that the rays are unchanged by N, for the initial ray has the equations

$$x'_\rho + \mathrm{i}x'_4\,\sigma'_\rho(\sigma'_\pi \sigma'_\pi + k'^2)^{-\frac{1}{2}} = 0,$$

and the corresponding final ray the equations

$$x_\rho + \mathrm{i}x_4\,\sigma'_\rho(\sigma'_\pi \sigma'_\pi + k^2)^{-\frac{1}{2}} = 0.$$

This example brings out a matter which might be a source of confusion. We have initial rays and final rays passing through a common event, viz. the space-time origin. Does this origin lie in M' or in M? It does not matter, because, since the rays are straight, we may conveniently produce them outside the domain to which they properly belong, just as in optics we produce rays behind a mirror in order to discuss a virtual image. This device is possible only for homogeneous media, since otherwise the rays are curved and there is no simple way of producing them outside their proper region.

Interpretation of T as action. In Hamilton's optics, T has a simple meaning in terms of optical length (cf. Synge, 1937, p. 29). The same sort of interpretation may be made in space-time.

Fig. 2.14 shows a ray proceeding from P' in M' to P in M. O is the origin in space-time. Since we restrict ourselves to homogeneous M' and M, the initial and final rays are straight, and σ'_r are constants along the initial ray, and σ_r constants along the final ray.

Let us consider the hyperplanes through O with the equations

$$\Pi': \ \sigma'_r y_r = 0; \quad \Pi: \sigma_r y_r = 0, \qquad (2.8.10)$$

where y_r are current coordinates. Let Q' and Q be the events at which Π' and Π cut the initial and final rays respectively, these rays being produced if necessary. Now we have seen that T is unchanged if we slide P' along the initial ray and P along the final ray, and we may even slide the events along the rays produced out of their proper regions, provided we make due allowance when doing so. Let us slide P' to Q' and P to Q. Then on account of (2.8.10) we may put $\sigma_r x_r = 0$, $\sigma'_r x'_r = 0$, and so obtain from (2.8.1) $V(Q', Q) = T$. Thus *we can express T as the action from Q' to Q*:

$$T = \int_{Q'}^{Q} f \, ds = [Q'Q], \quad (2.8.11)$$

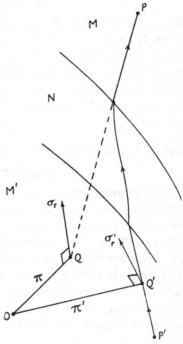

if we use the square brackets to indicate action. In the case of rays produced outside their proper region, the integral must be correctly interpreted. Thus, for fig. 2.14, we have

$$T = \int_{Q'}^{P} f \, ds - \int_{Q}^{P} f \, ds, \quad (2.8.12)$$

where in the first integral we use the 'natural' medium-function along the actual ray from Q' to P, but in the second we use the medium-function for M all along the straight (and partly artificial) ray from Q to P.

Fig. 2.14. Interpretation of T as action; $T = [Q'Q]$.

Calculation of T. Let us now see how T may be calculated. We consider only homogeneous regions, in each of which, as stated above, the medium-function is independent of position in space-time.

Consider a pair of regions, M', M, as in fig. 2.13, but let the intermediate region N shrink to a 3-space across which the medium-function f is discontinuous, so that we are back at the situation shown in fig. 2.11. Let y_r as before denote the coordinates of the event at which the ray from P' to P cuts N.

Now we have the expression (2.8.1) for T, and we know that the value of T is unchanged if we slide P' along the initial ray and P along the final ray. Let us slide them into coincidence at Q. Then

$$x'_r = x_r = y_r, \quad V(x', x) = 0,$$

and so (2.8.1) gives $\qquad T = (\sigma_r - \sigma_r') y_r.$ $\qquad\qquad$ (2.8.13)

Now N_r being normal to N at Q, we have from the law of refraction (2.7.6)

$$\frac{\sigma_\rho - \sigma_\rho'}{\sigma_4 - \sigma_4'} = \frac{N_\rho}{N_4}, \qquad\qquad (2.8.14)$$

and if we write the equation of N in the form

$$N(y) = i y_4 + S(y_1, y_2, y_3) = 0, \qquad\qquad (2.8.15)$$

then $\qquad\qquad N_\rho = \dfrac{\partial S}{\partial y_\rho} = S_\rho \text{ (say)}, \quad N_4 = i, \qquad\qquad (2.8.16)$

and (2.8.14) becomes $\qquad \dfrac{\sigma_\rho - \sigma_\rho'}{\sigma_4 - \sigma_4'} = -i S_\rho.$ $\qquad\qquad$ (2.8.17)

Then using the medium equations (2.8.7) and rewriting (2.8.13), we have

$$\left.\begin{array}{l} T = (\sigma_\rho - \sigma_\rho') y_\rho - [H(\sigma) - H'(\sigma')] \, S(y), \\[2mm] \dfrac{\sigma_\rho - \sigma_\rho'}{H(\sigma) - H'(\sigma')} = S_\rho(y). \end{array}\right\} \qquad (2.8.18)$$

Here σ, σ', y in the functional symbols stand for σ_μ, σ_μ', y_μ. Therefore *the function $T(\sigma', \sigma)$ for the pair of regions M', M, separated by the 3-space of discontinuity N, is to be found by eliminating the three quantities y_ρ from the four equations* (2.8.18). That is as far as we can carry the explicit calculations unless we are given the two medium-functions and the form of N.

The method can be extended to give T for any number of homogeneous regions, separated by 3-spaces of discontinuity, but as in the case of optics, the algebra of elimination becomes formidable. To deal with three regions, we turn back to fig. 2.13 and suppose that M', N and M are all homogeneous, with discontinuities on passing from M' into N and from N into M. The plan is to find the T-function by the preceding method, first for the pair of regions M', N and then for the pair N, M. Let these be respectively $T(M', N)$ and $T(N, M)$. Then we follow the optical procedure (cf. Synge, 1937, p. 35), writing for the T-function of the complete system (M' and M separated by N)

$$T(M', M) = T(M', N) + T(N, M), \qquad\qquad (2.8.19)$$

and eliminating the three slowness components of N by the condition that the expression on the right-hand side of (2.8.19) shall be

stationary with respect to arbitrary variations of them. The justification for this is the same as in the optical theory.

Example. Let M' and M be characterized by

$$\Omega'(\sigma') = \sigma'_r\sigma'_r + k'^2 = 0, \quad \sigma'_4 = i(\sigma'_\pi\sigma'_\pi + k'^2)^{\frac{1}{2}} = iH'(\sigma'),$$

$$\Omega(\sigma) = \sigma_r\sigma_r + k^2 = 0, \quad \sigma_4 = i(\sigma_\pi\sigma_\pi + k^2)^{\frac{1}{2}} = iH(\sigma),$$

and let the 3-space N separating them be the pseudoparaboloid

$$iy_4 + \tfrac{1}{2}a^{-1}y_\rho y_\rho = 0,$$

a being a constant. Then

$$S(y) = \tfrac{1}{2}a^{-1}y_\rho y_\rho, \quad S_\rho = a^{-1}y_\rho,$$

and the last of (2.8.18) gives

$$y_\rho = a\frac{\sigma_\rho - \sigma'_\rho}{H(\sigma) - H'(\sigma')},$$

and hence
$$S(y) = \tfrac{1}{2}a\frac{(\sigma_\rho - \sigma'_\rho)(\sigma_\rho - \sigma'_\rho)}{[H(\sigma) - H'(\sigma')]^2}.$$

Then by the first of (2.8.18) we have

$$T(\sigma', \sigma) = \tfrac{1}{2}a\frac{(\sigma_\rho - \sigma'_\rho)(\sigma_\rho - \sigma'_\rho)}{H(\sigma) - H'(\sigma)}.$$

Note that σ_ρ, σ'_ρ are six independent quantities here, and so we can
use (2.8.9) for the rays if we wish to investigate them. This independence is due to the fact that the normals to a pseudoparaboloid
have ∞^3 directions in space-time, and so, given σ'_r, we have three
degrees of freedom in σ_r, corresponding to the three degrees of
freedom in the position of the initial ray.

T for refraction at a fixed surface. There is a case of some importance which cannot be handled by the general method given above,
viz. *when the 3-space of discontinuity N is the history of a surface fixed
in the space of the Galilean observer.* We cannot then solve $N(y) = 0$
for y_4 as in (2.8.15); in fact, the equation of N is of the form

$$S(y_1, y_2, y_3) = 0. \tag{2.8.20}$$

To treat this case, we go back to (2.8.13), which is always valid.
The law of refraction (2.7.6) gives

$$\sigma_\rho - \sigma'_\rho = \theta S_\rho, \quad \sigma_4 - \sigma'_4 = 0, \tag{2.8.21}$$

θ being undetermined. Hence we can write the set of five equations

$$T = (\sigma_\rho - \sigma'_\rho) y_\rho, \quad \sigma_\rho - \sigma'_\rho = \theta S_\rho, \quad S(y) = 0, \quad (2.8.22)$$

and so obtain T by eliminating the four quantities y_ρ, θ.

Note that now σ_ρ, σ'_ρ are not independent, for they are connected by $H(\sigma) = H'(\sigma')$, and so only five of them are independent. Thus we cannot use (2.8.9) for the rays; let us see what the correct equations for the rays are.

We go back to (2.8.4), which reads

$$\delta T = x_\rho \delta \sigma_\rho + x_4 \delta \sigma_4 - x'_\rho \delta \sigma'_\rho - x'_4 \delta \sigma'_4,$$

or

$$\delta T = (x_\rho + i x_4 H_\rho) \delta \sigma_\rho - (x'_\rho + i x'_4 H'_\rho) \delta \sigma'_\rho, \quad (2.8.23)$$

where $H_\rho = \partial H / \partial \sigma_\rho$, $H'_\rho = \partial H' / \partial \sigma'_\rho$. The differentials in (2.8.23) are independent except for

$$H_\rho \delta \sigma_\rho - H'_\rho \delta \sigma'_\rho = 0, \quad (2.8.24)$$

and so, no matter how T is expressed in terms of σ'_ρ, σ_ρ (and it can, of course, be expressed in an infinite number of ways, since these quantities are not independent), we have

$$\left. \begin{array}{l} \dfrac{\partial T}{\partial \sigma'_\rho} = -x'_\rho - i x'_4 H'_\rho - \phi H'_\rho, \\[2mm] \dfrac{\partial T}{\partial \sigma_\rho} = x_\rho + i x_4 H_\rho + \phi H_\rho, \\[2mm] H'(\sigma') = H(\sigma), \end{array} \right\} \quad (2.8.25)$$

where ϕ is an undetermined multiplier.

These seven equations are the equations of the rays. Suppose we are given the slowness of the initial ray (σ'_ρ) and also an event on it (x'_r), so that the initial ray is completely known. Then we can solve the first and last of (2.8.25) for σ_ρ, ϕ, and when these are put in the second of (2.8.25) we have the three equations of the final ray corresponding to the given initial ray. Or, of course, we can use the equations in other ways.

T for refraction through a fixed hole. As a final case, let us calculate T for refraction through a hole fixed in the observer's space. Let us take the equations of the history of the hole to be

$$y_\rho = a_\rho, \quad (2.8.26)$$

so that a_ρ are the spatial coordinates of the hole. The formula (2.8.13) is a valid deduction from (2.8.1), and it now reads

$$T = (\sigma_\rho - \sigma_\rho')\, a_\rho + (\sigma_4 - \sigma_4')\, y_4, \qquad (2.8.27)$$

where $y_4 = ict$ on the hole. The appropriate form of the law of refraction is (2.7.14), and it gives

$$\sigma_4 - \sigma_4' = 0. \qquad (2.8.28)$$

Hence $\qquad\qquad T(\sigma', \sigma) = (\sigma_\rho - \sigma_\rho')\, a_\rho, \qquad (2.8.29)$

a very simple expression.

But here again σ_ρ, σ_ρ' are connected by (2.8.28), which gives $H'(\sigma') = H(\sigma)$, and we cannot use (2.8.9) for the rays. Actually they have the same form as (2.8.25), viz.

$$\left. \begin{aligned} -a_\rho &= -x_\rho' - ix_4' H_\rho' - \phi H_\rho', \\ a_\rho &= x_\rho + ix_4 H_\rho + \phi H_\rho, \\ H'(\sigma') &= H(\sigma). \end{aligned} \right\} \qquad (2.8.30)$$

We verify that the first line gives a straight line in space-time passing through the event $x_\rho' = a_\rho$, $x_4' = i\phi$ in the history of the hole, and the second line of (2.8.30) gives a straight line also passing through the same event.

2.9. Focal properties

The equations (2.8.9) are useful for the discussion of focal properties. Let us for simplicity assume that we are dealing with a congruence of rays in which the initial rays are parallel, i.e. the initial waves are plane. Then σ_ρ' are constants, and we work with

$$\frac{\partial T}{\partial \sigma_\rho} = x_\rho + ix_4 \frac{\partial H}{\partial \sigma_\rho}, \qquad (2.9.1)$$

in which σ_ρ' occur only as constants in the left-hand side. Here are three equations connecting the seven variables x_r, σ_ρ.

Apart from this restriction, the argument is general. The situation is as pictured in fig. 2.13, homogeneous regions, M', M, being separated by a region N which need not be homogeneous.

We now ask whether any final ray σ_ρ is approached to within the second order by any neighbouring ray, i.e. a ray with slowness $\sigma_\rho + \delta\sigma_\rho$. Any such point of approach we call a *focal event* (fig. 2.15).

We investigate the existence of focal events by applying a variation to (2.9.1) with $\delta x_r = 0$; this gives

$$T_{\pi\rho}\delta\sigma_\pi = \mathrm{i}x_4 H_{\pi\rho}\sigma_\pi, \qquad (2.9.2)$$

where the subscripts to T and H denote differentiation with respect to the σ's. These equations are consistent if, and only if,

$$\det(T_{\pi\rho} + ctH_{\pi\rho}) = 0, \qquad (2.9.3)$$

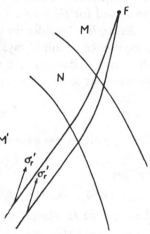

Fig. 2.15. Focal event F in space-time.

where we have put $x_4 = \mathrm{i}ct$, so that everything is real. To find focal events on an assigned final ray, we regard σ_ρ as known and solve (2.9.3) for t; when t has been found, we get the other coordinates of the focal event from (2.9.1).

Since (2.9.3) is a cubic equation, it must have one real root, and hence *every final ray has at least one focal event on it and not more than three.* However, it must be noted that we are considering the general case, for which (2.8.9) gives the rays, i.e. σ_ρ, σ'_ρ are independent quantities. The case where they are not independent (refraction at a fixed surface) will be discussed later in this section.

If the three roots of (2.9.3) are distinct, we speak of each of the focal events given by it as a *focal event of the first class*; if two roots coincide, we have a *focal event of the second class*; and if all three roots coincide, we have a *focal event of the third class*, or *principal focus*.

If at least one of the two quadratic forms

$$T_{\pi\rho}X_\pi X_\rho, \quad H_{\pi\rho}X_\pi X_\rho$$

is known to be positive-definite (or negative-definite), then the three roots of (2.9.3) are all equal if, and only if,

$$T_{\pi\rho} = \phi H_{\pi\rho}, \qquad (2.9.4)$$

where ϕ is a factor of proportionality. Thus if we seek the principal foci, not on one particular final ray, but in the whole congruence of final rays resulting from a given parallel initial congruence defined by σ'_ρ, we have to solve *six* equations (2.9.4) for *four* unknowns, σ_ρ and ϕ. Hence, in general, we are not to expect the existence of principal foci. This is different from what holds in ordinary optics; there principal foci do in general exist, in the sense that we have *three* equations of the form (2.9.4) (with suffixes in the range 1, 2) to be solved for *three* unknowns.

Example. Consider the congruence of rays formed from a parallel congruence of initial rays by refraction at a pseudoparaboloid, as discussed following (2.8.19). As the algebra is rather heavy in general, let us take for the initial rays $\sigma'_\rho = 0$, so that

$$T = \tfrac{1}{2}a\frac{\sigma_\rho\sigma_\rho}{H(\sigma)-H'(0)}.$$

Then, putting $a = 1$ for simplicity and writing

$$G(\sigma) = H(\sigma) - H'(0),$$

we get

$$T_{\rho\pi} = G^{-1}\delta_{\rho\pi} - G^{-2}(\sigma_\rho H_\pi + \sigma_\pi H_\rho) + G^{-3}\sigma_\mu\sigma_\mu H_\rho H_\pi - \tfrac{1}{2}\sigma_\mu\sigma_\mu H_{\rho\pi}.$$

Let us take $H(\sigma) = (\sigma_\pi\sigma_\pi + k^2)^{\frac{1}{2}}$, as earlier. Then

$$H_\rho = H^{-1}\sigma_\rho, \quad H_{\rho\pi} = H^{-1}\delta_{\rho\pi} - \sigma_\rho\sigma_\pi H^{-3},$$

and so

$$T_{\rho\pi} = A\delta_{\rho\pi} + B\sigma_\rho\sigma_\pi,$$

where A and B are easily calculated. Then (2.9.3) reads

$$\det[\delta_{\rho\pi}(A + ctH^{-1}) + \sigma_\rho\sigma_\pi(B - ctH^{-3})] = 0.$$

This is a cubic equation to determine the three possible focal points on the final ray σ_ρ. If we take in particular the final ray to be $\sigma_\rho = 0$, then $H = k$, and this determinantal equation becomes

$$(A + ctk^{-1})^3 = 0,$$

which has a triple root. Thus we get a principal focus at $ct = -Ak$, the other coordinates being $x_\rho = 0$, as is easily seen.

Focal events for refraction at a fixed surface. In the case of refraction at a fixed surface, as in (2.8.20), we have to proceed more circumspectly, because we can no longer use (2.8.9) for the rays; we must use (2.8.25) instead.

Let us again take the initial rays parallel ($\sigma'_\rho = $ const.); we have then

$$\frac{\partial T}{\partial\sigma_\rho} = x_\rho + (\mathrm{i}x_4 + \phi)H_\rho, \quad H(\sigma) = H'(\sigma'). \qquad (2.9.5)$$

To get a focal event, we hold x_r fixed and vary σ_ρ and ϕ. Thus

$$\left.\begin{aligned}
T_{\rho\pi}\delta\sigma_\pi &=(ix_4+\phi)H_{\rho\pi}\delta\sigma_\pi+H_\rho\delta\phi,\\
H_\pi\delta\sigma_\pi &=0.
\end{aligned}\right\} \tag{2.9.6}$$

Elimination of $\delta\sigma_\pi$ and $\delta\phi$ gives the fourth-order determinantal equation

$$\begin{vmatrix} T_{\rho\pi}+(ct-\phi)H_{\rho\pi} & H_\rho \\ H_\pi & 0 \end{vmatrix}=0. \tag{2.9.7}$$

Here σ_ρ are supposed given if we are investigating focal events on a given final ray. When (2.9.7) has been solved for $ct-\phi$ (or $ix_4+\phi$), substitution in (2.9.5) gives the coordinates x_ρ at the focal event. The value of t at the focal event remains undetermined, because there is a line of focal events parallel to the t-axis, occurring on parallel final rays which have all the same σ_ρ but different positions in space-time.

The above method is applied in §3.3.

2.10. Special types of medium-functions

In Chapter III we shall consider those medium-functions $f(x,\alpha)$ which appear to be of chief physical interest. Here, without taking physical significance into account, we shall discuss some types of medium-functions which deserve attention by virtue of special symmetries which they impart to regions of space-time.

The properties of a region with respect to rays and waves are summed up geometrically in the 3-spaces S and W of (2.1.19) and (2.1.20), S being the unit 3-wave and W the 3-surface of slowness. They are equivalent to one another on account of the relationship of polar reciprocity, and so it suffices to consider S only.

Isotropic regions. A region of space-time is *isotropic* at an event in it if the unit 3-wave S relative to that event looks the same to all Galilean observers (i.e. is invariant under Lorentz transformations). It is clear then that an isotropic region has a medium-function of the form

$$f(x,\alpha)=\phi(x)(-\alpha_r\alpha_r)^{\frac12}. \tag{2.10.1}$$

It is *homogeneous isotropic* if ϕ is a mere constant, and *heterogeneous isotropic* if ϕ varies with position in space-time.

We have already used this as an illustrative example. Let us sum up the properties of an isotropic region, these properties having already been established incidentally or being easy to prove.

The variational principle (2.1.3) now reads

$$\delta \int \phi(x)\, ds = 0, \tag{2.10.2}$$

and the equations (2.1.4) for the rays are

$$\frac{d}{ds}(\phi\alpha_r) + \phi_r = 0, \tag{2.10.3}$$

where $\phi_r = \partial\phi/\partial x_r$. We have

$$\sigma_r = -\frac{\partial f}{\partial \alpha_r} = \phi\alpha_r, \tag{2.10.4}$$

and so

$$\Omega(\sigma, x) = \tfrac{1}{2}(\sigma_r \sigma_r + \phi^2) = 0, \tag{2.10.5}$$

the factor $\tfrac{1}{2}$ being merely a matter of convenience. The Hamilton-Jacobi equation is

$$\frac{\partial V}{\partial x_r}\frac{\partial V}{\partial x_r} + \phi^2 = 0. \tag{2.10.6}$$

The Hamiltonian equations of a ray are, as in (2.1.17),

$$\frac{dx_r}{dw} = \sigma_r, \quad \frac{d\sigma_r}{dw} = -\phi\phi_r, \tag{2.10.7}$$

the parameter w being related to s by

$$f\frac{ds}{dw} = -\sigma_r\frac{\partial\Omega}{\partial\sigma_r} = -\sigma_r\sigma_r = \phi^2, \quad dw = \frac{fds}{\phi^2} = \frac{ds}{\phi}. \tag{2.10.8}$$

If we use w as parameter, the equations of a ray have the particular first integral

$$\frac{dx_r}{dw}\frac{dx_r}{dw} + \phi^2 = 0. \tag{2.10.9}$$

The unit 3-wave is the pseudosphere

$$S: \quad y_r y_r = -\phi^{-2}, \tag{2.10.10}$$

and the 3-surface of slowness W the reciprocal pseudosphere

$$W: \quad y_r y_r = -\phi^2. \tag{2.10.11}$$

As already shown in § 2.5, the wave velocity u_ρ is related to the ray velocity v_ρ by

$$u_\rho = v_\rho \frac{c^2}{v^2}, \quad uv = c^2. \tag{2.10.12}$$

Homogeneous regions. A *homogeneous* region of space-time is one for which $f(x, \alpha)$, and hence $\Omega(\sigma, x)$, is independent of x_r, so that we may write

$$f = f(\alpha). \tag{2.10.13}$$

Now the equations (2.1.4) for the rays give

$$\frac{\mathrm{d}}{\mathrm{d}s}\frac{\partial f}{\partial \alpha_r}=0, \quad \frac{\partial f}{\partial \alpha_r}=\text{const.}, \quad \alpha_r=\text{const.}, \qquad (2.10.14)$$

and so the rays are straight lines in space-time. This is the great simplification due to homogeneity; the integration of the equations of the rays is, in other cases, usually a matter of considerable difficulty.

If the region is both homogeneous and isotropic, then

$$f=k(-\alpha_r\alpha_r)^{\frac{1}{2}}, \qquad (2.10.15)$$

where k is a constant. Everything is then comparatively simple. Details will be worked out in §3.1 when we deal with the geometrical mechanics of a free particle.

We shall now consider some cases where the region is isotropic and possesses special symmetries so that the equations of the rays can be integrated.

Cases where the rays can be found easily. Consider, as a particular case of (2.10.1),

$$f(x,\alpha)=\phi(x_4)(-\alpha_r\alpha_r)^{\frac{1}{2}}, \qquad (2.10.16)$$

so that the unit 3-waves are pseudospheres with radii depending only on time for some Galilean observer. Then by (2.10.7) the rays satisfy

$$\frac{\mathrm{d}x_r}{\mathrm{d}w}=\sigma_r, \quad \frac{\mathrm{d}\sigma_\rho}{\mathrm{d}w}=0, \quad \frac{\mathrm{d}\sigma_4}{\mathrm{d}w}=-\phi\phi_4. \qquad (2.10.17)$$

Thus $\sigma_\rho=\text{const.}$ and $\quad x_\rho=\sigma_\rho w+a_\rho, \qquad (2.10.18)$

where a_ρ are constants. Further

$$\frac{\mathrm{d}x_4}{\mathrm{d}w}=\sigma_4=\mathrm{i}(\sigma_\pi\sigma_\pi+\phi^2)^{\frac{1}{2}}, \qquad (2.10.19)$$

and so the integration of the equations of the rays is completed by

$$\mathrm{i}w=\int(\sigma_\pi\sigma_\pi+\phi^2)^{-\frac{1}{2}}\,\mathrm{d}x_4. \qquad (2.10.20)$$

Another case in which the equations of the rays can be integrated explicitly to within a quadrature is

$$f(x,\alpha)=\phi(R)(-\alpha_r\alpha_r)^{\frac{1}{2}}, \quad R^2=-x_rx_r, \qquad (2.10.21)$$

this expression being available only inside the null-cone having the origin for vertex, since R is imaginary outside that cone. The unit

3-waves are pseudospheres, with radii depending only on R, and the theory is analogous to that of an isotropic optical medium with refractive index distributed with spherical symmetry. By (2.10.7) the equations of the rays are

$$\frac{\mathrm{d}x_r}{\mathrm{d}w} = \sigma_r, \quad \frac{\mathrm{d}\sigma_r}{\mathrm{d}w} = x_r \frac{\phi\phi'}{R}, \qquad (2.10.22)$$

where $\phi' = \mathrm{d}\phi/\mathrm{d}R$. Hence

$$\frac{\mathrm{d}^2 x_r}{\mathrm{d}w^2} = x_r \frac{\phi\phi'}{R}, \qquad (2.10.23)$$

and so we get 'angular momentum' integrals

$$x_r \frac{\mathrm{d}x_s}{\mathrm{d}w} - x_s \frac{\mathrm{d}x_r}{\mathrm{d}w} = C_{rs}, \qquad (2.10.24)$$

where C_{rs} are constants for the ray ($C_{rs} = -C_{sr}$).

To study any particular ray, we may use a special Galilean frame of reference, thus simplifying the skew-symmetric tensor C_{rs}. Let us in fact choose axes so that the x_4-axis passes through the initial event on the ray, and then choose the x_1-axis so that the initial direction is included in the 2-space of x_1 and x_4. This means that initially we have

$$x_1 = x_2 = x_3 = 0, \quad x_4 = iR, \quad \frac{\mathrm{d}x_2}{\mathrm{d}w} = \frac{\mathrm{d}x_3}{\mathrm{d}w} = 0. \qquad (2.10.25)$$

Now it is clear from (2.10.23) that if any coordinate and its first derivative both vanish initially, then that coordinate is permanently zero. Hence $x_2 = x_3 = 0$ along the ray, and we have from (2.10.24)

$$x_1 \frac{\mathrm{d}x_4}{\mathrm{d}w} - x_4 \frac{\mathrm{d}x_1}{\mathrm{d}w} = -iC, \quad R^2 = -x_1^2 - x_4^2, \qquad (2.10.26)$$

where C is a real constant. Let us now define θ by

$$x_1 = R\sinh\theta, \quad x_4 = iR\cosh\theta, \qquad (2.10.27)$$

so that $\theta = 0$ initially. Then (2.10.26) gives

$$R^2 \frac{\mathrm{d}\theta}{\mathrm{d}w} = C. \qquad (2.10.28)$$

Our problem is analogous to that of the Newtonian motion of a particle in a central field, and this equation is the analogue of the integral of angular momentum. The analogue of the equation of energy is (2.10.9), which becomes, by (2.10.27),

$$\left(\frac{dx_1}{dw}\right)^2 + \left(\frac{dx_4}{dw}\right)^2 = -\left(\frac{dR}{dw}\right)^2 + R^2\left(\frac{d\theta}{dw}\right)^2 = -\phi^2. \quad (2.10.29)$$

Putting $u = 1/R$, this equation, with (2.10.28), yields

$$\left(\frac{du}{d\theta}\right)^2 - u^2 = \frac{\phi^2}{C^2}, \quad (2.10.30)$$

which gives the relation between u and θ by a quadrature, and so gives us the ray. We verify easily that, if $\phi = $ const., then the ray is a straight line in space-time.

An interesting special choice of ϕ is

$$\phi(R) = \frac{k}{R}, \quad f(x,\alpha) = \frac{k}{R}(-\alpha_r\alpha_r)^{\frac{1}{2}}, \quad R^2 = -x_r x_r, \quad (2.10.31)$$

where k is a constant. Then (2.10.30) becomes

$$\left(\frac{du}{d\theta}\right)^2 = \left(1 + \frac{k^2}{C^2}\right)u^2, \quad (2.10.32)$$

and so

$$\frac{1}{u} = R = R_0 e^{K\theta}, \quad K = \pm\left(1 + \frac{k^2}{C^2}\right)^{\frac{1}{2}}, \quad (2.10.33)$$

where R_0 is the initial value of R. By (2.10.27) the ray has the equations

$$x_1 = R_0 e^{K\theta}\sinh\theta, \quad x_2 = x_3 = 0, \quad x_4/i = ct = R_0 e^{K\theta}\cosh\theta, \quad (2.10.34)$$

θ being a parameter on the ray. Let us choose the positive sense of the x_1-axis in the direction of motion at $\theta = 0$; then $K > 0$. Since

$$R^2 = c^2t^2 - x_1^2 = R_0^2 e^{2K\theta}, \quad \frac{ct}{x_1} = \coth\theta,$$

and the ray velocity is

$$v_1 = \frac{dx_1}{dt} = c\frac{K\tanh\theta + 1}{K + \tanh\theta} \quad (2.10.35)$$

(which is always less than c numerically), we have the following table of values for the ray:

θ	x_1	ct	R	ct/x_1	v_1	
$-\infty$	0	0	0	-1	$-c$	
0	0	R_0	R_0	∞	c/K	(2.10.36)
∞	∞	∞	∞	1	c	

The space-time diagram of the ray corresponding to $f(x, \alpha)$ as in (2.10.31) is shown in fig. 2.16. It starts from the origin with velocity $-c$, travelling first in the negative direction of x_1, reverses its direction of motion and arrives at the spatial origin again at time $t = R_0/c$ with velocity c/K and then goes off to $x_1 = \infty$ with a velocity increasing up to c as the limit.

Statical regions. Let us now consider a case of considerable interest, namely, the *statical* case in which $f(x, \alpha)$ is independent of

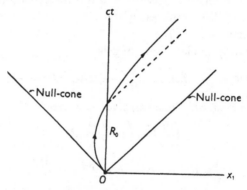

Fig. 2.16. Ray for $f(x, \alpha) = k(-\alpha_r \alpha_r)^{\frac{1}{2}}/R$.

time (i.e. independent of x_4) for some Galilean observer. Then $\Omega(\sigma, x)$ also is independent of x_4, and the equations (2.1.17) for the rays read

$$\frac{dx_\rho}{dw} = \frac{\partial \Omega}{\partial \sigma_\rho}, \quad \frac{d\sigma_\rho}{dw} = -\frac{\partial \Omega}{\partial x_\rho}, \qquad (2.10.37)$$

$$\frac{dx_4}{dw} = \frac{\partial \Omega}{\partial \sigma_4}, \quad \sigma_4 = \text{const.} \qquad (2.10.38)$$

We note that (2.10.37) involve only six dependent variables, x_ρ, σ_ρ, since x_4 does not appear and σ_4 is a constant by (2.10.38). Thus in this case the problem of finding the rays with given σ_4 as curves in space is precisely a problem in Hamiltonian optics in Euclidean 3-space based on the relation

$$\Omega(\sigma_1, \sigma_2, \sigma_3, \sigma_4, x_1, x_2, x_3) = 0, \qquad (2.10.39)$$

with σ_4 an assigned constant, corresponding to the colour of the light. This means that if we were interested only in statical regions, separated by fixed surfaces of discontinuity (and, indeed, the cases of most obvious physical interest are of this type), then we might scrap the space-time mode of thought and fall back on the classical optics of Hamilton. If we did this, however, we would lose the simple interpretation of de Broglie waves as 3-waves in space-time, and so it seems better to keep the wider view which embraces non-statical cases.

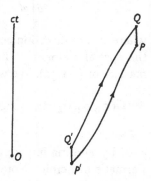

Fig. 2.17. Statical case;
$$\partial f / \partial x_4 = 0.$$

Space-time diagram of variation of ray in Fermat's principle ($\delta \sigma_4 = 0$; $\delta x_\rho = 0$ at ends).

We shall now establish Fermat's principle, or a principle equivalent to his, for a statical region. Consider a ray $P'P$ in space-time (fig. 2.17), and let us vary this to an arbitrary neighbouring curve $Q'Q$, the terminal infinitesimal displacements $P'Q'$, PQ being parallel to the t-axis. Let us associate with the events on $Q'Q$ values of σ_r satisfying

$$\Omega(\sigma, x) = 0, \quad \sigma_4 = b_4, \quad (2.10.40)$$

the values of σ_r on $Q'Q$ being close to those on $P'P$, and b_4 being the constant value of σ_4 on $P'P$. In this variation we have

$$\delta \int \sigma_r \, dx_r = \int (\delta \sigma_r \, dx_r + \sigma_r \, \delta dx_r)$$

$$= \left[\sigma_4 \delta x_4 \right]_{P'}^{P} + \int (\delta \sigma_r \, dx_r - \delta x_r \, d\sigma_r). \quad (2.10.41)$$

Since $P'P$ is a ray, it follows from (2.1.17) that the last integral is $\int \delta \Omega \, dw$, and this vanishes by (2.10.40). Further

$$\delta \int \sigma_4 \, dx_4 = \sigma_4 \delta \int dx_4 = \left[\sigma_4 \delta x_4 \right]_{P'}^{P}. \quad (2.10.42)$$

Then, subtracting (2.10.42) from (2.10.41), we have

$$\delta \int \sigma_\rho \, dx_\rho = 0, \quad (2.10.43)$$

where (2.10.40) is satisfied on the varied curve and at the ends $\delta x_\rho = 0$. All trace of time has disappeared from (2.10.43), and so we have a variational principle which we can apply in space, σ_4 being of course assigned. To pass to a principle of Fermat's form

$$\delta \int g(x_1, x_2, x_3, \beta_1, \beta_2, \beta_3) \, ds_0 = 0, \qquad (2.10.44)$$

where β_ρ are direction cosines of a space curve $(\beta_\rho \beta_\rho = 1)$ and ds_0 the spatial element of length $(ds_0^2 = dx_\rho \, dx_\rho)$, we proceed in the manner of (2.1.33). We write the five equations

$$\Omega(\sigma_1, \sigma_2, \sigma_3, \sigma_4, x_1, x_2, x_3) = 0, \quad \beta_\rho = \theta \frac{\partial \Omega}{\partial \sigma_\rho}, \quad g = +\theta \sigma_\rho \frac{\partial \Omega}{\partial \sigma_\rho},$$
$$(2.10.45)$$

and, by eliminating σ_ρ and θ, obtain g, the refractive index in Fermat's principle. Note that the last sign in (2.10.45) is positive, and not negative as in (2.1.33); this is because we are dealing with space instead of space-time, and the line element is positive-definite.

The analogue of the optical instrument of revolution. The instrument of revolution plays such an important part in geometrical optics that it is natural to ask what its analogue in geometrical mechanics might be. The most obvious analogue is a set of space-time regions, each homogeneous and isotropic, separated by 3-spaces of discontinuity each of which has an equation involving only x_4 and $x_\rho x_\rho$, just as the surfaces in an instrument of revolution have equations involving only z and $x^2 + y^2$. Any such 3-space is then the history of an expanding or contracting sphere with centre at the origin of space. The t-axis takes over the role of the axis of symmetry of the instrument of revolution. The paraxial rays of optics become rays with small velocities.

Such a system is specified by writing down a set of equations for 3-surfaces of slowness of the form

$$\Omega(\sigma, x) = \sigma_r \sigma_r + k^2 = 0, \qquad (2.10.46)$$

the constant k taking different values in the different regions, and a set of equations for the separating 3-spaces, in general of the form (cf. (2.8.15))

$$i y_4 + S(y_\rho y_\rho) = 0, \qquad (2.10.47)$$

or, if the 3-space is fixed in the observer's space (cf. (2.8.20)),

$$y_\rho y_\rho = \text{const.} \qquad (2.10.48)$$

This last would correspond in the optical instrument to a circular cylinder having for axis the axis of the instrument.

Before making any calculations at all, it is clear from symmetry that the T-function of the complete system, considered as a function of the three initial slowness components σ'_ρ and the three final slowness components σ_ρ, can contain these variables only in the forms

$$\sigma'_\rho \sigma'_\rho, \quad \sigma_\rho \sigma'_\rho, \quad \sigma_\rho \sigma_\rho, \tag{2.10.49}$$

these being the only invariants under those Lorentz transformations which leave the t-axis unchanged.

As regards the calculation of $T(\sigma', \sigma)$, in the work following (2.8.19) we have as an example calculated T for refraction across a pseudoparaboloid. This work will serve, in general, for the calculation of T up to the second order inclusive for small σ'_ρ, σ_ρ, since then the rays will be near the t-axis and (2.10.47) will have as leading terms

$$iy_4 + \tfrac{1}{2}a^{-1}y_\rho y_\rho = 0, \tag{2.10.50}$$

if we choose the origin of space-time on the 3-space. Such results can be combined in the manner of (2.8.19) to obtain the T-function for the complete system. To this order of accuracy it would give the 'Gaussian geometrical mechanics' of the set of regions.

CHAPTER III

GEOMETRICAL MECHANICS FOR A PARTICLE, FREE OR IN A GIVEN FIELD

3.1. Basic formulae for a free particle

Just as in optics we first study the simplest type of medium (homogeneous and isotropic), so we should first study the geometrical mechanics of a free particle.

For a free particle of proper mass m, we take the medium-function

$$f(x, \alpha) = mc(-\alpha_r \alpha_r)^{\frac{1}{2}}, \qquad (3.1.1)$$

which will be recognized as homogeneous and isotropic. The constant factor mc gives the integral (2.1.5) the dimensions of action.

The rays satisfy

$$\delta \int f(x, \alpha) \, ds = mc \, \delta \int ds = 0,$$

and so are straight lines in space-time, the rays through an event x_r' having the equations

$$x_r = x_r' + \alpha_r s,$$

where α_r are constants for the ray ($\alpha_r \alpha_r = -1$) and s the Minkowskian distance measured from x_r'. This corresponds to uniform motion with velocity v_ρ given by

$$\alpha_\rho = \gamma v_\rho / c, \quad \alpha_4 = i\gamma, \quad \gamma = (1 - v^2/c^2)^{-\frac{1}{2}}, \quad v^2 = v_\rho v_\rho, \quad (3.1.2)$$

as in (2.4.6); α_r is the velocity 4-vector of the particle.

The characteristic function is

$$\begin{aligned} V(x', x) &= mc[-(x_r - x_r')(x_r - x_r')]^{\frac{1}{2}} \\ &= mc(c^2\tau^2 - r^2)^{\frac{1}{2}}, \end{aligned} \qquad (3.1.3)$$

where τ is the interval of time and r the spatial distance between the events x_r' and x_r. The slowness 4-vector is, by (2.1.9),

$$\sigma_r = -\frac{\partial f}{\partial \alpha_r} = mc\alpha_r(-\alpha_n \alpha_n)^{-\frac{1}{2}} = mc\alpha_r, \qquad (3.1.4)$$

so that, by (3.1.2), $\qquad \sigma_\rho = m\gamma v_\rho, \quad \sigma_4 = im\gamma c. \qquad (3.1.5)$

Thus for a free particle the slowness 4-vector is, in fact, the usual momentum-energy 4-vector.

The coincidence in space-time direction of the slowness 4-vector σ_r and the tangent vector to the ray α_r is a property peculiar to isotropic medium-functions; it holds, as we have just seen, for a free particle, but it does not hold in general for a particle moving in an electromagnetic field. The congruence of space-time curves defined by the directions of σ_r for a system of waves is *always* a normal congruence, cutting the 3-waves normally, but in general the congruence of rays is not a normal congruence, i.e. there are no 3-spaces cut normally by the rays. But in the case of a free particle the rays do form a normal congruence, on account of (3.1.4).

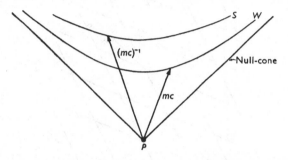

Fig. 3.1. Unit 3-wave S of radius $(mc)^{-1}$ and 3-surface of slowness W of radius mc, for a free particle.

Elimination of the α's from (3.1.4) by the relation $\alpha_r \alpha_r = -1$ gives immediately the slowness equation (2.1.14) in the form

$$\Omega(\sigma, x) = \sigma_r \sigma_r + m^2 c^2 = 0, \qquad (3.1.6)$$

and the Hamilton-Jacobi equation is

$$\frac{\partial V}{\partial x_r} \frac{\partial V}{\partial x_r} + m^2 c^2 = 0. \qquad (3.1.7)$$

The unit 3-wave of (2.1.19) and the 3-surface of slowness (2.1.20) are now

$$\begin{aligned} S: \quad & mc(-y_r y_r)^{\frac{1}{2}} = 1, \\ W: \quad & y_r y_r + m^2 c^2 = 0. \end{aligned} \qquad (3.1.8)$$

Thus S is a pseudosphere of radius $(mc)^{-1}$ and W a pseudosphere of radius mc (fig. 3.1). The kinematical interpretation of S given in (2.4.10) now shows us an expanding sphere of flashing lamps carried by particles emitted simultaneously from a point, in all directions and with all speeds up to c, the flashing time τ of the lamp carried by a particle with speed v being given by

$$mc\tau(c^2 - v^2)^{\frac{1}{2}} = 1. \qquad (3.1.9)$$

Since the medium-function is independent of x_r, the unit 3-wave is a member of the family of 3-waves from a source-event; this family will be discussed below (cf. (3.1.13)).

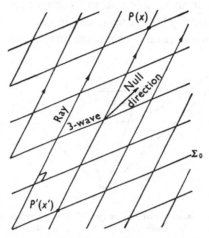

Fig. 3.2. Space-time diagram of straight rays and plane 3-waves for a free particle. The rays are normal to the 3-waves (in the Minkowskian sense).

The general relationship (2.5.11), expressing wave velocity in terms of the slowness 4-vector, gives, by (3.1.4),

$$u_\rho = -\,\mathrm{i}c\,\frac{\sigma_\rho\sigma_4}{\sigma_\pi\sigma_\pi} = -\,\mathrm{i}c\,\frac{\alpha_\rho\alpha_4}{\alpha_\pi\alpha_\pi} = -\,\mathrm{i}c\,\frac{v_\rho\mathrm{i}c}{v_\pi v_\pi} = \frac{c^2}{v^2}v_\rho. \qquad (3.1.10)$$

Thus for a free particle the directions of the wave velocity u_ρ and the ray (or particle) velocity v_ρ coincide, and their magnitudes are connected by the de Broglie relationship

$$uv = c^2. \qquad (3.1.11)$$

Plane waves and waves from a source-event. Of all the manifold systems of waves which can be associated with the motion of a free particle, two are particularly simple and interesting: (a) plane waves, (b) waves from a source-event.

Let us start from an initial plane 3-wave Σ_0. On it σ_r lies along the normal (general property), and (3.1.4) tells us that α_r does so too (special property for free particle). Thus we get a congruence in space-time of straight parallel rays, the orthogonal trajectories of a system of parallel plane 3-waves (fig. 3.2). The kinematical picture, obtained by slicing across the space-time diagram by hyperplanes $t = \text{const.}$, shows us parallel plane 2-waves advancing with velocity u_ρ; the particle velocity v_ρ is normal to the 2-waves and the relationship $uv = c^2$ is satisfied (fig. 3.3).

The characteristic function for this set of plane waves is (cf. fig. 3.2)

$$V(P) = mcP'P = -mc\alpha_r(x_r - x_r') = -\sigma_r x_r + \text{const.,} \quad (3.1.12)$$

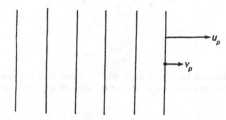

Fig. 3.3.   Kinematical picture in space of plane 2-waves advancing with velocity u_ρ in the direction of the ray (or particle) velocity v_ρ ($uv = c^2$).

since $\alpha_r x_r'$ is a constant for any choice of P' on Σ_0. V is a function of the coordinates of P only, as is always the case when we are dealing with the congruence of rays generated by an initial 3-wave. In (3.1.12) σ_r are the constant components of the slowness 4-vector, which, as we have seen, is also the momentum-energy 4-vector.

Consider now the waves from a source-event $P'(x')$. The 3-waves are found by putting $V = \text{const.}$ in (3.1.3), and so their equations are

$$c^2\tau^2 - r^2 = (V/mc)^2 = \text{const.} \quad (3.1.13)$$

These are pseudospheres as shown in fig. 3.4. The corresponding 2-waves are expanding spheres, the wave velocity being

$$u = \frac{dr}{d\tau} = \frac{c^2\tau}{r}. \quad (3.1.14)$$

Fig. 3.5 shows a selection of such 2-waves at time τ, namely, those with velocities c, $2c$, $3c$, $4c$, $5c$. The first of these is a sphere with radius $c\tau$; it bounds the waves at time τ, and strictly speaking does not belong to the system, since it corresponds to a particle velocity c.

It must be remembered that there is no wave-length or frequency in this geometrical mechanics, and if certain waves are picked out in the diagrams, they are to be regarded as samples picked out in an arbitrary way. There are no crests or troughs or loci of equal phase,

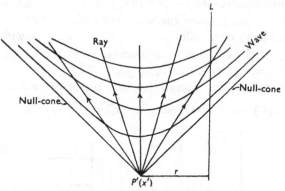

Fig. 3.4. Space-time diagram of rays and 3-waves from a source-event $P'(x')$ for a free particle. L is the world line of a fixed observer.

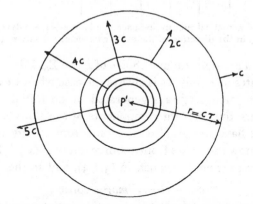

Fig. 3.5. Kinematical picture of 2-waves from a source-event P' (at time τ after that event), showing the 2-waves with velocities $c, 2c, 3c, 4c, 5c$, the first being the bounding 2-wave with radius $r=c\tau$. The velocity is inversely proportional to the radius.

because the theory contains no phase. Phase is a matter of quantization and will not be introduced until Chapter IV.

Focal mirrors in space-time. In ordinary optics it is easy to bring to a real focus P all the rays emerging from a source P'. All we have to do is to construct a mirror of the form $r'+r=$ const., where r', r

are distances from P', P respectively. The mirror is an ellipsoid of revolution having P', P for foci, the two meanings of the word *focus* (geometrical and optical) coinciding in this case.

What is the analogue of this focal mirror in the geometrical mechanics of a free particle? Given two events, P' and P, is it possible to construct a mirror in space-time so that all the rays from P' (a source-event) will pass through P after reflexion, the rays

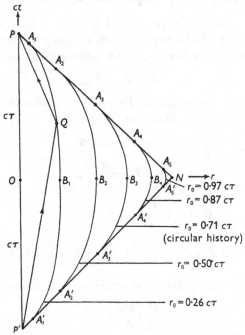

Fig. 3.6. Space-time histories of expanding and contracting spherical mirrors reflecting a source-event P' into an event P. (Ellipsoidal 3-mirror.)

corresponding to the motions of free particles and reflexion taking place according to the law of stationary action?

It is obvious that it cannot be done if the space-time displacement $P'P$ is space-like or null. Let us consider the case where $P'P$ is time-like.

We may apply a Lorentz transformation so as to make $P'P$ the time-axis, and then we can choose the origin of space-time half-way between these events, so that the times at P' and P are $-\tau$ and τ respectively (fig. 3.6).

Let Q be an event with coordinates r, ct, where r is the spatial distance of Q from the origin, i.e. from the common position of P' and P. (In space-time diagrams it is convenient to plot ct instead of t, so that the null lines make angles of $45°$ with the axes.) We choose Q so that $P'Q$ and QP are both time-like and point into the future. Let us write R', R for the Minkowskian distances $P'Q$, QP, so that

$$R' = [c^2(t+\tau)^2 - r^2]^{\frac{1}{2}}, \quad R = [c^2(t-\tau)^2 - r^2]^{\frac{1}{2}}. \quad (3.1.15)$$

Since the action for the path $P'QP$ is $mc(R'+R)$, the variational principle (2.1.3) tells us any ray from P' will pass through P after reflexion in a 3-space (let us call it a 3-*mirror*) with equation

$$R' + R = 2K, \quad (3.1.16)$$

where K is a positive constant. To get the equation of this 3-mirror in convenient form, we note that by (3.1.15)

$$R'^2 - R^2 = 4ct.c\tau, \quad (3.1.17)$$

and hence

$$R' - R = \frac{2ct.c\tau}{K},$$

$$R' = K + \frac{ct.c\tau}{K}, \quad R = K - \frac{ct.c\tau}{K}. \quad (3.1.18)$$

Therefore

$$c^2t^2 + c^2\tau^2 - r^2 = \tfrac{1}{2}(R^2 + R'^2) = K^2 + \frac{(ct)^2(c\tau)^2}{K^2}, \quad (3.1.19)$$

and so the equation of the focal 3-mirror takes the simple form

$$\frac{r^2}{r_0^2} + \frac{c^2t^2}{K^2} = 1, \quad (3.1.20)$$

where

$$r_0 = (c^2\tau^2 - K^2)^{\frac{1}{2}}. \quad (3.1.21)$$

We choose K in the range $0 < K < c\tau$.

According to (3.1.20) the 3-mirror is, in a sense, an ellipsoid of revolution in space-time, but that geometrical description is appropriate only to the Euclidean metric of the paper on which the space-time diagram is drawn. For a given pair of events, P' and P, there is a single infinity of focal 3-mirrors depending on the choice of K, the semi-axes being r_0 and K.

The null-cone drawn into the past from P has the equation

$$r = c(\tau - t), \quad (3.1.22)$$

and the intersection of this cone with the 3-mirror (3.1.20) is given by

$$(rc\tau - r_0^2)^2 = 0; \qquad (3.1.23)$$

this shows that the 3-mirror *touches* the null-cone at

$$r = \frac{r_0^2}{c\tau}, \quad ct = \frac{K^2}{c\tau}. \qquad (3.1.24)$$

Fig. 3.6 shows a selection of focal 3-mirrors drawn for various values of r_0 (equivalently, various values of K). These are represented by the elliptical arcs $A_1' B_1 A_1, \ldots, A_5' B_5 A_5$, these arcs touching the null lines $P'N$, NP. Only the arcs between the events of tangency are shown, because only these parts are operative; if the other parts of the ellipses were used as mirrors, the reflected rays would go into the past, which is not allowed.

In ordinary optics we can trap *all* the rays from a point-source and bring them to a point-focus. But in space-time some of the rays are bound to escape, e.g. the rays drawn from P' to $A_1 P$ in the case of the 3-mirror $A_1' B_1 A_1$. To trap as many rays as possible, we should make r_0 as small as possible, but we trap them all only in the limit $r_0 = 0$.

Viewed kinematically, the focal 3-mirror is the history of a sphere which expands and contracts. The smallest radius is $r_0^2/c\tau$, and then the points of the sphere are moving with velocity c. The radius expands from this minimum to a maximum r_0, and then contracts again to the minimum radius, with velocity c inwards. What it does after that is of no importance, provided its velocity does not exceed c, for it is no longer acting as a focal mirror; it is only catching the lost rays.

As regards the acceleration of the mirror, we have by (3.1.20)

$$\left. \begin{array}{l} r^2 = r_0^2 \left(1 - \dfrac{c^2 t^2}{K^2}\right), \quad r \dfrac{\mathrm{d}r}{\mathrm{d}t} = -\dfrac{r_0^2 c^2 t}{K^2}, \\[2mm] r \dfrac{\mathrm{d}^2 r}{\mathrm{d}t^2} + \left(\dfrac{\mathrm{d}r}{\mathrm{d}t}\right)^2 = -\dfrac{r_0^2 c^2}{K^2}, \end{array} \right\} \qquad (3.1.25)$$

and hence

$$\frac{\mathrm{d}^2 r}{\mathrm{d}t^2} = -\frac{c^2 r_0^4}{K^2 r^3}. \qquad (3.1.26)$$

Thus the radial acceleration varies as the inverse cube of the radius.

Just as we may use a small piece of an ellipsoidal mirror to reflect to a focus a small bundle of light rays from a point-source, so we may use a small portion of a focal 3-mirror to reflect to P a wave packet from a source-event P'. If we wish to catch in this way the packet containing the ray $P'B_1$ (fig. 3.6), all we need is to provide at time $t = 0$ (and a little before and after) a spherical mirror of radius r_0, instantaneously at rest, but with a radial acceleration

$$\frac{\mathrm{d}^2 r}{\mathrm{d}t^2} = -\frac{c^2 r_0}{c^2 \tau^2 - r_0^2}. \qquad (3.1.27)$$

Since the ray velocity is $v = r_0/\tau$, we may say that *a wave packet from a source-event, with ray (or group) velocity v, will be brought to simultaneous focus again at the position of the source if reflected in a spherical mirror of any radius r_0, provided that the mirror, when struck by the packet, is instantaneously at rest and has an acceleration of radial contraction of amount*

$$-\frac{\mathrm{d}^2 r}{\mathrm{d}t^2} = \frac{1}{r_0} \frac{v^2}{1 - v^2/c^2}. \qquad (3.1.28)$$

The paraboloidal 3-mirror. The same plan is also available to reflect rays from a source-event into a parallel beam, or to reflect a parallel beam so as to make it pass through a single event. This is the analogue of the paraboloidal mirror of optics.

Consider a source-event P'. We wish to set up a 3-mirror which will reflect all the rays from P' into a parallel beam, with an assigned space-time direction. Choose P' for origin and the assigned direction of the reflected beam for that of the time-axis (fig. 3.7). Then, if r is spatial distance from P', the action along a path $P'QP$ is

$$mc[(c^2t^2 - r^2)^{\frac{1}{2}} + c(\tau - t)], \qquad (3.1.29)$$

where (r, ct) are the coordinates of Q and τ is the time at P. If we are to get a set of rays parallel to the time-axis (equivalently, 3-waves $\tau = $ const.) as a result of reflexion in a mirror at Q, then (by the variational principle) the 3-mirror must satisfy

$$(c^2t^2 - r^2)^{\frac{1}{2}} - ct = -a, \qquad (3.1.30)$$

where a is a constant. Thus *the equation of the paraboloidal 3-mirror is*

$$r^2 - 2act + a^2 = 0. \qquad (3.1.31)$$

This touches the null-cone $P'N$ at

$$r = ct = a. \qquad (3.1.32)$$

The paraboloidal 3-mirror is shown as AQB in fig. 3.7. It traps *all* the rays from the source-event P' and makes them parallel to the time-axis; in other words, *it reduces to rest all particles shot out simultaneously from P' in all directions and with all velocities less than* c. The parameter a may have any positive value.

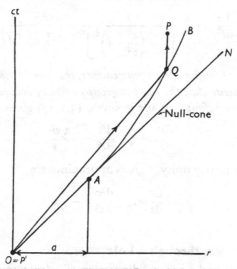

Fig. 3.7. Space-time history AQB of expanding spherical mirror, reflecting rays from the source-event P' into a set of parallel rays, i.e. reducing particles to rest. (Paraboloidal 3-mirror.)

Kinematically, the paraboloidal 3-mirror is the history of an expanding sphere, the radius at time t being given by (3.1.31). The velocity of expansion is

$$\frac{\mathrm{d}r}{\mathrm{d}t} = c\frac{a}{r}, \tag{3.1.33}$$

and the acceleration

$$\frac{\mathrm{d}^2r}{\mathrm{d}t^2} = -\frac{c^2a^2}{r^3}. \tag{3.1.34}$$

Suppose now that we wish to catch only a wave packet with velocity v, using an expanding sphere which at the instant of reflexion has a radius r. We have $v = r/t$, and so (3.1.31) gives a in terms of r by the equation

$$r^2 - 2acr/v + a^2 = 0, \tag{3.1.35}$$

so that

$$\frac{r}{a} = \frac{c}{v} + \left(\frac{c^2}{v^2} - 1\right)^{\frac{1}{2}}, \tag{3.1.36}$$

the positive sign being chosen to make $r > a$. We get for the velocity and acceleration of expansion of the mirror at the instant of reflexion

$$\left.\begin{aligned}
\frac{\mathrm{d}r}{\mathrm{d}t} &= \frac{c}{\dfrac{c}{v}+\left(\dfrac{c^2}{v^2}-\mathrm{I}\right)^{\frac{1}{2}}} = \frac{v\gamma}{\mathrm{I}+\gamma}, \\
\frac{\mathrm{d}^2r}{\mathrm{d}t^2} &= -\frac{c^2}{r}\,\frac{\mathrm{I}}{\left[\dfrac{c}{v}+\left(\dfrac{c^2}{v^2}-\mathrm{I}\right)^{\frac{1}{2}}\right]^2} = -\frac{v^2}{r}\,\frac{\gamma^2}{(\mathrm{I}+\gamma)^2}.
\end{aligned}\right\} \tag{3.1.37}$$

If the mirror satisfies these requirements, the wave packet is reduced to rest, in the sense that the ray (or group) velocity becomes zero and the wave velocity infinite. If v/c is small, (3.1.37) gives approximately

$$\frac{\mathrm{d}r}{\mathrm{d}t} = \tfrac{1}{2}v, \quad \frac{\mathrm{d}^2r}{\mathrm{d}t^2} = -\frac{\mathrm{I}}{4}\frac{v^2}{r}, \tag{3.1.38}$$

and if v/c is nearly unity it gives approximately

$$\frac{\mathrm{d}r}{\mathrm{d}t} = c, \quad \frac{\mathrm{d}^2r}{\mathrm{d}t^2} = -\frac{c^2}{r}. \tag{3.1.39}$$

3.2. Refraction through a hole

In § 2.7 we had a general discussion of refraction through a hole in a screen, and in (2.8.29) an expression was given for $T(\sigma', \sigma)$ in the case where the two regions separated by the screen were homogeneous. We shall now give a direct determination of the characteristic function V for the refraction of a free particle through a fixed hole.

In fig. 3.8 the straight line H parallel to the time-axis represents the history of the hole, and M' and M are the regions of space-time separated by the history of the screen. Let $P'(x')$ and $P(x)$ be events in M' and M respectively, and let the ray from P' to P meet H at $Q(y)$. Then the characteristic function is

$$V(P', P) = mc(R' + R), \tag{3.2.1}$$

where
$$\begin{aligned}
R'^2 &= P'Q^2 = -(x'_r - y_r)(x'_r - y_r), \\
R^2 &= PQ^2 = -(x_r - y_r)(x_r - y_r).
\end{aligned} \tag{3.2.2}$$

The quantities y_r are to be eliminated from the expression (3.2.1) by the condition (principle of stationary action) that its value shall be stationary with respect to variation of y_r on H. Differentiation with respect to y_4 gives the condition

$$\frac{x_4'-y_4}{R'}+\frac{x_4-y_4}{R}=0. \tag{3.2.3}$$

Let us choose the origin of space-time on H, so that $y_\rho=0$, and let us write
$$r'^2=x_\rho'x_\rho', \quad r^2=x_\rho x_\rho, \quad y_4=ic\tau, \tag{3.2.4}$$

so that r' and r are the spatial distances of P' and P respectively from the hole and τ is the time at Q. We have in this notation

$$\left.\begin{array}{l}R'^2=c^2(\tau-t')^2-r'^2,\\ R^2=c^2(t-\tau)^2-r^2.\end{array}\right\} \tag{3.2.5}$$

Now (3.2.3) gives

$$\left.\begin{array}{l}c(\tau-t')=\theta R',\\ c(t-\tau)=\theta R,\end{array}\right\} \tag{3.2.6}$$

where θ is a factor of proportionality, and when we substitute in (3.2.5) we get

$$\left.\begin{array}{l}R'^2(\theta^2-1)=r'^2,\\ R^2(\theta^2-1)=r^2.\end{array}\right\} \tag{3.2.7}$$

Elimination of τ from (3.2.6) gives

$$c(t-t')=\theta(R'+R)=(r'+r)\theta(\theta^2-1)^{-\frac{1}{2}}, \tag{3.2.8}$$

and so θ is expressed in terms of the end variables by

$$\theta=c(t-t')[c^2(t-t')^2-(r'+r)^2]^{-\frac{1}{2}}. \tag{3.2.9}$$

Hence by (3.2.8)
$$R'+R=[c^2(t-t')^2-(r'+r)^2]^{\frac{1}{2}}, \tag{3.2.10}$$

and so (3.2.1) gives as *the characteristic function for the refraction of a free particle through a fixed hole*

$$V(P',P)=mc[c^2(t-t')^2-(r'+r)^2]^{\frac{1}{2}}, \tag{3.2.11}$$

where t' and t are the times of the initial and final events and r', r their distances from the hole. This may be checked by considering the particular ray which goes straight through the hole, in which case $r'+r$ is the spatial distance from P' to P; indeed it would not be

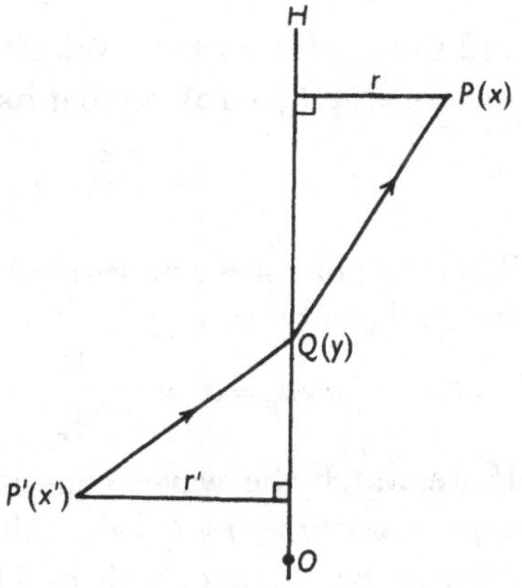

Fig. 3.8. Space-time diagram of the refraction of a ray from a source-event $P'(x')$ through a hole H

hard to construct a proof of (3.2.11) in general by starting with this particular case.

By (3.2.11) the 3-waves in M due to the source-event $P'(x')$ in M' are given by

$$(r+r')^2 = c^2(t-t')^2 - (V/mc)^2, \qquad (3.2.12)$$

where V, r', t' are constants. The 2-waves are given by putting $t = \text{const.}$, so that each 2-wave is a sphere with centre at the hole; the wave velocity is

$$u = \frac{dr}{dt} = \frac{c^2(t-t')}{r+r'}, \qquad (3.2.13)$$

and the ray (or particle) velocity in M is then given by the general relationship (3.1.10), so that its magnitude is

$$v = \frac{r+r'}{t-t'}. \qquad (3.2.14)$$

This may of course be verified directly by differentiating (3.2.11), for we have by (3.1.5)

$$m\gamma v_\rho = \sigma_\rho = -\frac{\partial V}{\partial x_\rho}, \quad im\gamma c = \sigma_4 = -\frac{\partial V}{\partial x_4} = \frac{i}{c}\frac{\partial V}{\partial t}. \qquad (3.2.15)$$

If we watch the waves coming out of the hole ($r=0$), they first appear at time $t = t' + r'/c$, the wave velocity and the particle velocity being then both c. The wave velocity at the hole then increases linearly to infinity, the particle velocity decreasing to zero.

Refraction of plane 3-waves incident on a hole. Consider the refraction through a fixed hole with world line H (fig. 3.9) of plane incident 3-waves with constant slowness 4-vector σ_r'. The characteristic function is as in (3.2.11), but from that expression we have to eliminate r' and t' by the condition that it shall be stationary with respect to variations of x_r' subject to

$$\sigma_r' x_r' = \text{const.}, \qquad (3.2.16)$$

this being the equation of some initial 3-wave Σ_0.

It is convenient to work with real quantities, and so we write (3.2.16) as (cf. (3.2.15))

$$v_\rho' x_\rho' - c^2 t' = C, \qquad (3.2.17)$$

v_ρ' being the incident ray velocity. Then, varying x_ρ' and t' in (3.2.11) and (3.2.17), we get

$$\left. \begin{aligned} c^2(t-t')\,\delta t' + (r+r')\,\delta r' &= 0, \\ v_\rho'\,\delta x_\rho' - c^2 \delta t' &= 0, \end{aligned} \right\} \qquad (3.2.18)$$

the former to hold for all variations which satisfy the latter. Taking the hole at $y_\rho = a_\rho$, we have

$$r'^2 = (x'_\rho - a_\rho)(x'_\rho - a_\rho), \quad r' \delta r' = (x'_\rho - a_\rho) \delta x'_\rho,$$

and so we get from (3.2.18)

$$(r+r')(x'_\rho - a_\rho) = -\phi v'_\rho, \qquad (3.2.19)$$

$$r'(t-t') = \phi, \qquad (3.2.20)$$

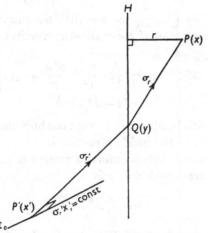

Fig. 3.9. Space-time diagram for refraction of plane 3-waves incident on a hole, H.

where ϕ is a factor of proportionality. We have now in (3.2.17), (3.2.19) and (3.2.20) five equations for the determination of x'_ρ, t' and ϕ.

By (3.2.19)

$$x'_\rho = a_\rho - \phi \frac{v'_\rho}{r+r'},$$

and so

$$\left. \begin{array}{l} r' = \phi \dfrac{v'}{r+r'}, \\[2mm] v'_\rho x'_\rho = v'_\rho a_\rho - \phi \dfrac{v'^2}{r+r'} \\[2mm] \qquad = v'_\rho a_\rho - v'r'. \end{array} \right\} \qquad (3.2.21)$$

Thus by (3.2.17), (3.2.20) we have the following two equations for r' and t':

$$v'_\rho a_\rho - v'r' - c^2 t' = C, \quad \frac{v'(t-t')}{r+r'} = 1. \qquad (3.2.22)$$

Solving, we find

$$r+r'=\frac{v'/c^2}{1-v'^2/c^2}(C+c^2t-v'r-v'_\rho a_\rho).\qquad(3.2.23)$$

Now by (3.2.11)

$$V(P)=mc(r+r')\left[\frac{c^2(t-t')^2}{(r+r')^2}-1\right]^{\frac{1}{2}},\qquad(3.2.24)$$

or by (3.2.22) $$V(P)=mc(r+r')\frac{c}{v'}\left(1-\frac{v'^2}{c^2}\right)^{\frac{1}{2}},\qquad(3.2.25)$$

and so finally by (3.2.23) we see that *the characteristic function V(P) for the refraction of plane incident 3-waves* (3.2.17) *through a fixed hole at* $y_\rho=a_\rho$ *is*

$$\left.\begin{array}{c}V(P)=mc^2\gamma'\left(t-\dfrac{rv'}{c^2}-\dfrac{v'_\rho a_\rho}{c^2}+\dfrac{C}{c^2}\right),\\[2mm]\gamma'=(1-v'^2/c^2)^{-\frac{1}{2}}.\end{array}\right\}\qquad(3.2.26)$$

Here r is the spatial distance of P from the hole and t the time of P; v'_ρ is the velocity of the incident particles.

The 2-waves from the hole form a system of spheres centred at the hole, and the wave velocity is

$$u=\frac{dr}{dt}=\frac{c^2}{v'}=\frac{c^2}{v},\qquad(3.2.27)$$

where v is the magnitude of the final particle velocity, unchanged by passage through the hole (conservation of energy).

The effect of replacing a source-event by plane incident waves is this: for the source-event the magnitude of the wave velocity behind the screen varies with position and time according to (3.2.13), whereas, as for plane incident waves, it is a constant, being in fact that of the incident waves.

3.3. Refraction at a potential-jump

Consider a medium-function

$$f(x,\alpha)=mc(-\alpha_r\alpha_r)^{\frac{1}{2}}-iU\alpha_4/c,\qquad(3.3.1)$$

where $U=U(x_1,x_2,x_3)$. Since x_4 is absent from $f(x,\alpha)$, we have the statical case, and the Euler-Lagrange equations (2.1.4) have the first integral

$$\sigma_4=-\frac{\partial f}{\partial\alpha_4}=mc\alpha_4+\frac{iU}{c}=\frac{i}{c}(m\gamma c^2+U)=i\frac{E}{c},\qquad(3.3.2)$$

where E is a constant which we recognize as the constant of energy. The other Euler-Lagrange equations give

$$\frac{d}{ds}\frac{\partial f}{\partial \alpha_\rho} + \frac{i\alpha_4}{c}\frac{\partial U}{\partial x_\rho} = 0. \qquad (3.3.3)$$

Since $\alpha_4 = i\gamma$, $\alpha_\rho = \gamma v_\rho/c$, these may be written

$$\frac{d}{ds}(m\gamma v_\rho) + \frac{\gamma}{c}\frac{\partial U}{\partial x_\rho} = 0, \qquad (3.3.4)$$

or, since $\gamma\,ds = c\,dt$, $\qquad \dfrac{d}{dt}(m\gamma v_\rho) = -\dfrac{\partial U}{\partial x_\rho}. \qquad (3.3.5)$

We recognize U as *potential energy*.

Suppose now that space-time is divided into two regions, M' and M, by a 3-space N, as in fig. 2.11. Let us assign to these regions the medium-functions

$$\begin{aligned} M': && f'(x',\alpha') &= mc(-\alpha_r'\alpha_r')^{\frac{1}{2}} - iU'\alpha_4'/c, \\ M: && f(x,\alpha) &= mc(-\alpha_r\alpha_r)^{\frac{1}{2}} - iU\alpha_4/c, \end{aligned} \qquad (3.3.6)$$

where U' and U are *constants*. Then N is a 3-space which represents the history of a *potential-jump*; we are going to investigate refraction across it.

We have
$$\begin{aligned} \sigma_\rho' &= -\frac{\partial f'}{\partial \alpha_\rho'} = mc\alpha_\rho', & \sigma_4' &= mc\alpha_4' + iU'/c, \\ \sigma_\rho &= -\frac{\partial f}{\partial \alpha_\rho} = mc\alpha_\rho, & \sigma_4 &= mc\alpha_4 + iU/c, \end{aligned} \qquad (3.3.7)$$

and so the slowness equations $\Omega' = 0$, $\Omega = 0$ of (2.1.14) read

$$\begin{aligned} \Omega'(\sigma') &= \sigma_\rho'\sigma_\rho' + (\sigma_4' - iU'/c)^2 + m^2c^2 = 0, \\ \Omega(\sigma) &= \sigma_\rho\sigma_\rho + (\sigma_4 - iU/c)^2 + m^2c^2 = 0. \end{aligned} \qquad (3.3.8)$$

Thus, as in (2.8.8),

$$\begin{aligned} \sigma_4' &= iH'(\sigma'), & H'(\sigma') &= \frac{U'}{c} + (m^2c^2 + \sigma_\rho'\sigma_\rho')^{\frac{1}{2}} = \frac{E'}{c}, \\ \sigma_4 &= iH(\sigma), & H(\sigma) &= \frac{U}{c} + (m^2c^2 + \sigma_\rho\sigma_\rho)^{\frac{1}{2}} = \frac{E}{c}, \end{aligned} \qquad (3.3.9)$$

where E' is energy before, and E the energy after, crossing the potential-jump.

Note that each of the equations (3.3.8) represents a complete pseudosphere of two sheets, the radius in each case being mc and the centres being $(0, 0, 0, iU'/c)$ and $(0, 0, 0, iU/c)$. But only the sheets given by (3.3.9) are admissible, positive values of the square roots being understood here and throughout. We see this from (3.3.7), in view of the fact that the 4-vectors α'_r and α_r must point into the future, so that α'_4/i and α_4/i must be positive.

Let us now suppose the potential-jump to be fixed in the observer's space, so that the equation of N is

$$S(y_1, y_2, y_3) = 0.$$

Then the law of refraction (2.8.21) may be written

$$\sigma_\rho - \sigma'_\rho = \theta n_\rho, \quad \sigma_4 - \sigma'_4 = 0, \qquad (3.3.10)$$

where θ is an undetermined factor and n_ρ is the unit normal to N, drawn in the sense of incidence, so that $n_\rho \sigma'_\rho > 0$. We note that by (3.3.7)

$$-ic\sigma'_4 = m\gamma'c^2 + U' = E', \quad -ic\sigma_4 = m\gamma c^2 + U = E; \quad (3.3.11)$$

thus the last of (3.3.10) expresses *conservation of energy*:

$$E = m\gamma c^2 + U = m\gamma'c^2 + U' = E'. \qquad (3.3.12)$$

The conditions (3.3.10) may be stated in this form:

(i) the change in momentum is normal to the surface of the potential-jump;

(ii) energy is conserved.

To find the refracted slowness σ_ρ corresponding to given incident slowness σ'_ρ, we have to find σ_ρ and θ from the four equations

$$\sigma_\rho = \sigma'_\rho + \theta n_\rho, \quad H(\sigma) = H'(\sigma'). \qquad (3.3.13)$$

Substitution for σ_ρ from the first in the last gives the following equation for θ:

$$\frac{U}{c} + [m^2c^2 + (\sigma'_\rho + \theta n_\rho)(\sigma'_\rho + \theta n_\rho)]^{\frac{1}{2}} = H'(\sigma'), \qquad (3.3.14)$$

in which, as always, the *positive* square root is understood. It is convenient to introduce the notation

$$\left. \begin{array}{l} W' = (m^2c^2 + \sigma'_\rho \sigma'_\rho)^{\frac{1}{2}} = m\gamma'c, \\ W = (m^2c^2 + \sigma_\rho \sigma_\rho)^{\frac{1}{2}} = m\gamma c, \\ \Sigma = [(\sigma_\rho - \sigma'_\rho)(\sigma_\rho - \sigma'_\rho)]^{\frac{1}{2}}, \end{array} \right\} \qquad (3.3.15)$$

so that, by (3.3.9),

$$E' = cH'(\sigma') = U' + cW', \quad E = cH(\sigma) = U + cW. \quad (3.3.16)$$

Then (3.3.14) may be written

$$[m^2c^2 + (\sigma'_\rho + \theta n_\rho)(\sigma'_\rho + \theta n_\rho)]^{\frac{1}{2}} = c^{-1}(cW' + U' - U).$$

$$(3.3.17)$$

The left-hand side is not less than mc, and so a *necessary* condition for refraction to take place is

$$E' = U' + cW' \geqslant U + mc^2. \qquad (3.3.18)$$

Thus, *for refraction to be possible, the incident particle must have more energy than a particle of the same proper mass at rest on the other side of the potential-jump.*

We now square both sides of (3.3.17) and obtain for θ the quadratic equation

$$\theta^2 + 2n_\rho \sigma'_\rho \theta - c^{-2}(U' - U)(2cW' + U' - U) = 0. \quad (3.3.19)$$

The condition for reality of roots is

$$(n_\rho \sigma'_\rho)^2 + c^{-2}(U' - U)(2cW' + U' - U) \geqslant 0. \qquad (3.3.20)$$

Conditions (3.3.18) and (3.3.20) are certainly *necessary* for refraction. We shall now show that they are *sufficient*. Suppose them satisfied. Then (3.3.19) has two real roots, θ_1 and θ_2, which satisfy (3.3.17), and so refracted rays are given by

$$\sigma_\rho^{(1)} = \sigma'_\rho + \theta_1 n_\rho, \quad \sigma_\rho^{(2)} = \sigma'_\rho + \theta_2 n_\rho,$$

satisfying (3.3.13). Then

$$n_\rho \sigma_\rho^{(1)} + n_\rho \sigma_\rho^{(2)} = 2n_\rho \sigma'_\rho + \theta_1 + \theta_2 = 0, \qquad (3.3.21)$$

since the sum of the roots of (3.3.19) is $-2n_\rho \sigma'_\rho$. Now, to pass into M, the refracted ray must satisfy $n_\rho \sigma_\rho > 0$. We see from (3.3.21) that one ray satisfies this and the other does not; the ray that satisfies it corresponds to the greater root of (3.3.19); the other ray must be rejected as an extraneous solution.

We see then that (3.3.18) and (3.3.20) are *necessary and sufficient* conditions for refraction to take place and that, when it does take place, there is a unique refracted ray.

We note that if $U < U'$, then (3.3.18) and (3.3.20) are both satisfied; *refraction always takes place if there is lower potential energy on the far side of the jump (potential hole).*

If $U > U'$, then (3.3.18) sets a lower bound for the magnitude of σ'_ρ if refraction is to take place; in fact we must have

$$c^2 \sigma'_\rho \sigma'_\rho \geqslant (U - U')^2 + 2mc^2(U - U'). \qquad (3.3.22)$$

Assuming that some value has been assigned to $\sigma'_\rho \sigma'_\rho$ satisfying this, (3.3.20) tells us that the incident ray cannot deviate too far from the normal to the surface of the jump if refraction is to take place; in fact, if i is the angle of incidence, we must have

$$\cos^2 i \geqslant \frac{(U - U')(2cW' + U' - U)}{\sigma'_\rho \sigma'_\rho}. \qquad (3.3.23)$$

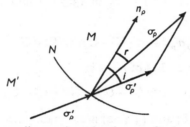

Fig. 3.10. Space diagram for refraction at a fixed potential-jump.

To consider refraction more intuitively, we may use a space diagram (fig. 3.10). Here M' and M are the two regions of space and N the potential-jump separating them. We have by (3.3.7) and (3.1.2)

$$\left.\begin{aligned}
\sigma'_\rho &= mc\alpha'_\rho = m\gamma' v'_\rho, \quad \gamma' = \left(1 - \frac{v'^2}{c^2}\right)^{-\frac{1}{2}}, \\
\sigma_\rho &= mc\alpha_\rho = m\gamma v_\rho, \quad \gamma = \left(1 - \frac{v^2}{c^2}\right)^{-\frac{1}{2}},
\end{aligned}\right\} \qquad (3.3.24)$$

where v'_ρ is the incident velocity and v_ρ the final velocity. Now $\sigma_\rho - \sigma'_\rho$ is normal to N, and so the angle of incidence i and the angle of refraction r satisfy

$$\frac{\sin i}{\sin r} = \frac{(\sigma_\rho \sigma_\rho)^{\frac{1}{2}}}{(\sigma'_\pi \sigma'_\pi)^{\frac{1}{2}}} = \frac{\gamma v}{\gamma' v'}, \qquad (3.3.25)$$

or, since $v^2/c^2 = 1 - \gamma^{-2}$, $v^2 \gamma^2 = c^2(\gamma^2 - 1)$,

$$\frac{\sin i}{\sin r} = \frac{(\gamma^2 - 1)^{\frac{1}{2}}}{(\gamma'^2 - 1)^{\frac{1}{2}}}. \qquad (3.3.26)$$

Thus $\gamma v/c$ or $(\gamma^2 - 1)^{\frac{1}{2}}$ plays the role of *refractive index* (more of this later). The relationship between γ' and γ is, as in (3.3.12),

$$m\gamma c^2 + U = m\gamma' c^2 + U'.$$

To study *reflexion*, we have instead of (3.3.13)

$$\sigma_\rho = \sigma'_\rho + \theta n_\rho, \quad H'(\sigma) = H'(\sigma'), \qquad (3.3.27)$$

since the final ray goes back into M'. Thus we are to write U' for U in (3.3.19), and so we get, as equation for θ,

$$\theta^2 + 2n_\rho \sigma'_\rho \theta = 0. \qquad (3.3.28)$$

The solution $\theta = 0$ is to be rejected as extraneous because it gives $\sigma_\rho = \sigma'_\rho$ and so makes the final ray go on into M. We have then

$$\theta = -2n_\rho \sigma'_\rho,$$

and so by (3.3.27) the reflected slowness 4-vector is

$$\sigma_\rho = \sigma'_\rho - 2n_\pi \sigma'_\pi n_\rho, \quad \sigma_4 = \sigma'_4. \qquad (3.3.29)$$

Reflexion is in fact very simple. The particle bounces with unchanged speed in a direction which satisfies the usual law of reflexion, viz. coplanarity of lines of incidence and reflexion with the normal and equality of the angles of incidence and reflexion.

Refraction at a fixed potential-jump may also be studied by means of a space-time diagram as in fig. 2.12. In fig. 3.11 the straight line N represents the history of the point on the surface at which the ray strikes, and Q is the event of striking. C is the simultaneous event in the history of the space-origin, so that $CQ = y_\rho$. The incident slowness 4-vector is $QA'_r = \sigma'_r$. Ω is the 3-space $\Omega = 0$, or rather the sheet $\sigma_4 = iH(\sigma)$ of (3.3.9). It is a pseudosphere of radius mc with centre at D where $QD = U/c$.

The condition $\sigma_4 = \sigma'_4$ of (3.3.10) tells us that we are to pass from the end of σ'_r to the end of σ_r (these 4-vectors being drawn from Q) by moving in a space-time direction which is orthogonal to the time-axis; the other three conditions in (3.3.10) tell us the precise direction in which we are to go. It is clear then, from the hyperboloidal character of the pseudosphere when drawn on Euclidean paper, that when we pass from A' (the end of σ'_r) in the assigned direction, we shall get two intersections with Ω or none (we omit the exceptional case of one intersection). A necessary condition for intersection with Ω (and therefore for refraction, since the end of σ_r must lie on Ω) is evidently

$$\sigma_4'/i > mc + \frac{U}{c},$$

or equivalently by (3.3.11)

$$m\gamma'c^2 + U' > mc^2 + U, \qquad (3.3.30)$$

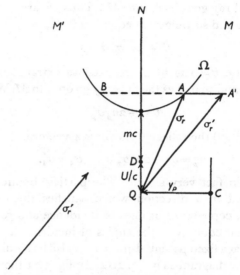

Fig. 3.11. Space-time diagram for refraction at a fixed potential-jump N, showing the slowness 4-vectors for the case $U > U'$.

which is the same as (3.3.18). Total reflexion occurs if

$$m\gamma'c^2 + U' < mc^2 + U. \qquad (3.3.31)$$

If the energy of the incident particle satisfies this inequality, it cannot cross the potential-jump, no matter what its direction of motion may be. We must remember that (3.3.30) is only a *necessary* condition for refraction; even though the particle has enough energy, it may not happen to cross the jump because (3.3.20) is not satisfied.

Fig. 3.11 shows two intersections, A and B, with the line drawn through A' in the appropriate direction. In order to cross N, $\sigma_\rho n_\rho$ must be positive, and it appears from the figure that B is extraneous, $\sigma_r = QA$ giving the unique refracted slowness 4-vector.

The fact that A' has been drawn outside Ω means that $U > U'$. If $U < U'$, A' would lie inside Ω and there would necessarily be two intersections, one giving an extraneous solution.

Fermat's principle. At this point it is interesting to go back to the medium-function (3.3.1), in which U is not a constant but some function of x_1, x_2, x_3. We have then a statical case, as considered at (2.10.39), and we can obtain a Fermat principle (2.10.44), using (2.10.45); in the present case (2.10.45) reads, if we put $\frac{1}{2}\theta$ for θ,

$$\left.\begin{array}{c} \sigma_\rho \sigma_\rho + \left(\sigma_4 - \dfrac{iU}{c}\right)^2 + m^2c^2 = 0, \\[2mm] \beta_\rho = \theta\sigma_\rho, \quad g = \theta\sigma_\rho\sigma_\rho. \end{array}\right\} \qquad (3.3.32)$$

Here g is the 'refractive index' of (2.10.44) and β_ρ are direction cosines, so that $\beta_\rho\beta_\rho = 1$, and we obtain

$$\left.\begin{array}{l} \theta = (\sigma_\rho\sigma_\rho)^{-\frac{1}{2}}, \\[1mm] g = (\sigma_\rho\sigma_\rho)^{\frac{1}{2}} \\[1mm] \quad = \left[-m^2c^2 - \left(\sigma_4 - \dfrac{iU}{c}\right)^2\right]^{\frac{1}{2}} \\[1mm] \quad = [-m^2c^2 - m^2c^2\alpha_4^2]^{\frac{1}{2}} \\[1mm] \quad = mc(\gamma^2 - 1)^{\frac{1}{2}}. \end{array}\right\} \qquad (3.3.33)$$

Now for a particle in a statical region with

$$f(x, \alpha) = mc(-\alpha_r\alpha_r)^{\frac{1}{2}} - iU(x_1, x_2, x_3)\,\alpha_4/c, \qquad (3.3.34)$$

a ray possesses the integral (cf. (3.3.2))

$$m\gamma c^2 + U = E. \qquad (3.3.35)$$

Any ray in space, when compared with adjacent curves traversed with that γ (or equivalently that v) determined by (3.3.35), satisfies Fermat's principle in the form $\delta\int g\,ds_0 = 0$ or

$$\delta\int(\gamma^2 - 1)^{\frac{1}{2}}\,ds_0 = 0, \qquad (3.3.36)$$

where ds_0 is the spatial line element $(ds_0^2 = dx_\rho\,dx_\rho)$. *Thus the refractive index is* $(\gamma^2 - 1)^{\frac{1}{2}}$, *as we already saw at* (3.3.26).

For slow motions $\gamma^2 - 1 = v^2/c^2$ approximately, and the variational principle (3.3.36) becomes approximately the principle of Maupertuis,

$$\delta\int v\,ds_0 = 0, \qquad (3.3.37)$$

where, by (3.3.35) to the same approximation, v is determined by

$$\tfrac{1}{2}mv^2 + U = \text{const.} \tag{3.3.38}$$

T for refraction at a fixed spherical potential-jump. Let us now return to the case of two homogeneous regions, M' and M, as in (3.3.6), separated by a potential-jump N, and discuss the calculation of $T(\sigma', \sigma)$ for refraction across the jump. The plan for this is already laid down in (2.8.22); we have to eliminate y_ρ and θ from the five equations

$$T = (\sigma_\rho - \sigma'_\rho)y_\rho, \quad \sigma_\rho - \sigma'_\rho = \theta S_\rho, \quad S(y) = 0. \tag{3.3.39}$$

We cannot proceed further without choosing the shape of the potential-jump. Let us take it to be *spherical* with radius a, so that

$$S(y) = \tfrac{1}{2}(y_\rho y_\rho - a^2) = 0. \tag{3.3.40}$$

Equations (3.3.39) now read

$$T = (\sigma_\rho - \sigma'_\rho)y_\rho, \quad \sigma_\rho - \sigma'_\rho = \theta y_\rho, \quad y_\rho y_\rho = a^2, \tag{3.3.41}$$

and we find at once

$$a\theta = \epsilon[(\sigma_\rho - \sigma'_\rho)(\sigma_\rho - \sigma'_\rho)]^{\frac{1}{2}}, \tag{3.3.42}$$

where $\epsilon = \pm 1$, the sign to be determined later. It follows that

$$T(\sigma', \sigma) = \epsilon a[(\sigma_\rho - \sigma'_\rho)(\sigma_\rho - \sigma'_\rho)]^{\frac{1}{2}}; \tag{3.3.43}$$

this is the T-function for refraction across a fixed spherical potential-jump with equation (3.3.40).

The sign of ϵ is determined as follows. We note that ϵ has the same sign as T and therefore the same sign as $(\sigma_\rho - \sigma'_\rho)y_\rho$. Now define ϵ_1 and ϵ_2 as follows:

$\epsilon_1 = +1$ if $U > U'$,

$\epsilon_1 = -1$ if $U < U'$,

$\epsilon_2 = +1$ if the incident ray strikes a convex surface of the sphere,

$\epsilon_2 = -1$ if the incident ray strikes a concave surface of the sphere.

A change in sign of ϵ_1 changes the signs of the following:

$U' - U$,

$\gamma' - \gamma$ by (3.3.12),

$i - r$ by (3.3.26),

$(\sigma_\rho - \sigma'_\rho)y_\rho$ by fig. 3.10,

ϵ.

On the other hand, a change in the sign of ϵ_2 leaves fig. 3.10 unchanged except for the form of N, but it changes the sign of y_ρ and so changes the sign of ϵ. Hence $\epsilon = \pm\,\epsilon_1\epsilon_2$, and it is easily seen that in fig. 3.10 we have

$$i > r, \quad \gamma > \gamma', \quad U < U', \quad \epsilon_1 = -1, \quad \epsilon_2 = +1,$$
$$(\sigma_\rho - \sigma'_\rho)\,y_\rho < 0, \quad \epsilon = -1,$$

and so the correct formula for the determination of ϵ is

$$\epsilon = \epsilon_1\epsilon_2, \tag{3.3.44}$$

where ϵ_1 and ϵ_2 are defined as above.

It is a usual thing to find an ambiguity of sign occurring in the calculation of a T-function, and then the ambiguity has to be removed by the consideration of special cases, as above.

Having now got T, we seek the equations for the rays. On account of $H(\sigma) = H'(\sigma')$, the quantities σ_ρ, σ'_ρ are not independent, and we have to use (2.8.25) for the rays instead of (2.8.9). Thus the equations of the rays are

$$\left. \begin{aligned} \epsilon a(\sigma'_\rho - \sigma_\rho)/\Sigma &= -x'_\rho + (ct' - \phi)\,\sigma'_\rho/W', \\ \epsilon a(\sigma_\rho - \sigma'_\rho)/\Sigma &= x_\rho - (ct - \phi)\,\sigma_\rho/W, \\ U + cW &= U' + cW', \end{aligned} \right\} \tag{3.3.45}$$

in the notation of (3.3.15).

Foci for refraction at a fixed spherical potential-jump. It seems a good practice to apply general methods to particular problems for which the solution can be obtained, perhaps more easily, in other ways; by such applications we come to understand what the general methods mean. Hamilton left the simple applications of his optical method unpublished, and to that fact we must attribute the long neglect which his method suffered. It would be unwise to repeat this mistake in carrying over his method into relativistic geometrical mechanics.

In this spirit let us study, by means of the T-function, the focal properties of the rays formed from a system of parallel rays by refraction at a fixed spherical potential-jump, recognizing that if we wished we could at once translate this problem into the terms of ordinary optics by means of the law of refraction in the form (3.3.26), or equivalently by the Fermat principle (3.3.36). Viewed in space-time, our problem is analogous to that of finding the caustic surfaces formed by refraction in a cylindrical rod of glass, for although the potential-jump is spherical in space, its history in space-time is cylindrical.

For the determination of foci we have, as in (2.9.6),

$$\left.\begin{array}{l} T_{\rho\pi}\delta\sigma_\pi = (\phi - ct)H_{\rho\pi}\delta\sigma_\pi + H_\rho\delta\phi, \\ H_\pi\delta\sigma_\pi = 0, \end{array}\right\} \quad (3.3.46)$$

where, as in (3.3.9),

$$H = \frac{U}{c} + (m^2c^2 + \sigma_\rho\sigma_\rho)^{\frac{1}{2}}. \quad (3.3.47)$$

We substitute for T from (3.3.43) and get, in the notation of (3.3.15),

$$\left.\begin{array}{l} \dfrac{\epsilon a}{\Sigma}\delta\sigma_\rho - \dfrac{\epsilon a}{\Sigma^3}(\sigma_\rho - \sigma'_\rho)(\sigma_\pi - \sigma'_\pi)\delta\sigma_\pi \\ \qquad = (\phi - ct)\left(\dfrac{\delta\sigma_\rho}{W} - \dfrac{\sigma_\rho\sigma_\pi\delta\sigma_\pi}{W^3}\right) + \dfrac{\sigma_\rho}{W}\delta\phi, \\ \sigma_\pi\delta\sigma_\pi = 0. \end{array}\right\} \quad (3.3.48)$$

The *caustic 3-surfaces* (loci of focal events) are to be found by eliminating $\delta\sigma$ and $\delta\phi$ from these four homogeneous equations, and so obtaining the determinantal equation

$$\begin{vmatrix} \dfrac{\epsilon a}{\Sigma^3}[\delta_{\rho\pi}\Sigma^2 - (\sigma_\rho - \sigma'_\rho)(\sigma_\pi - \sigma'_\pi)] + \dfrac{ct - \phi}{W^3}[\delta_{\rho\pi}W^2 - \sigma_\rho\sigma_\pi], & \dfrac{\sigma_\rho}{W} \\ \sigma_\pi, & 0 \end{vmatrix} = 0.$$
$$(3.3.49)$$

With this we associate four equations from (3.3.45), viz.

$$\frac{\epsilon a}{\Sigma}(\sigma_\rho - \sigma'_\rho) = x_\rho - \frac{ct - \phi}{W}\sigma_\rho, \quad U + cW = U' + cW'. \quad (3.3.50)$$

Elimination of σ_ρ and $ct - \phi$ from (3.3.49) and (3.3.50) leaves us with an equation involving x_ρ and σ'_ρ. But we are dealing with parallel incident rays and so σ'_ρ are given constants. Thus the equation in question gives us a 3-space, which is in fact the caustic 3-surface; it is a cylinder with generators parallel to the time-axis. Needless to say, eliminations of this sort are easier to speak of than to carry out!

Let us now look for focal events on that particular final ray for which σ_ρ are proportional to σ'_ρ, so that

$$\sigma_\rho = \beta\sigma'_\rho, \quad (3.3.51)$$

where β is a factor of proportionality. We find its value from the last of (3.3.50), which gives

$$\frac{U}{c} + (m^2c^2 + \beta^2\sigma'_\rho\sigma'_\rho)^{\frac{1}{2}} = \frac{U'}{c} + W', \qquad (3.3.52)$$

and so
$$\beta = (\sigma'_\rho\sigma'_\rho)^{-\frac{1}{2}}\left[\left(\frac{U'-U}{c} + W'\right)^2 - m^2c^2\right]^{\frac{1}{2}}. \qquad (3.3.53)$$

Let us now choose, as of course we may, the x_1-axis in the direction of σ'_ρ, so that $\sigma'_1 > 0$, $\sigma'_2 = 0$, $\sigma'_3 = 0$. Then the four equations (3.3.48) read

$$\begin{aligned}
\frac{\epsilon a}{\Sigma}\delta\sigma_1 - \frac{\epsilon a}{\Sigma^3}(\beta-1)^2\sigma_1'^2\delta\sigma_1 &= (\phi-ct)\left(\frac{1}{W} - \frac{\beta^2\sigma_1'^2}{W^3}\right)\delta\sigma_1 + \frac{\beta\sigma'_1}{W}\delta\phi, \\
\frac{\epsilon a}{\Sigma}\delta\sigma_2 = \frac{\phi-ct}{W}\delta\sigma_2, \quad & \frac{\epsilon a}{\Sigma}\delta\sigma_3 = \frac{\phi-ct}{W}\delta\sigma_3, \quad \sigma'_1\delta\sigma_1 = 0.
\end{aligned}$$
$$(3.3.54)$$

These are satisfied by $\delta\sigma_1 = 0$, $\delta\phi = 0$ and arbitrary values of $\delta\sigma_2$ and $\delta\sigma_3$, provided that we have

$$\phi - ct = \frac{\epsilon a}{\Sigma}W. \qquad (3.3.55)$$

We have then a focal event of the second class; to locate it, we go back to (3.3.50) and note that, by (3.3.15),

$$\Sigma^2 = (1-\beta)^2\sigma'_\rho\sigma'_\rho = (1-\beta)^2\sigma_1'^2, \quad W = (m^2c^2 + \beta^2\sigma'_\rho\sigma'_\rho)^{\frac{1}{2}}. \qquad (3.3.56)$$

Thus the focal event is at

$$x_\rho = \frac{\epsilon a}{\Sigma}(\beta-1)\sigma'_\rho - \frac{\epsilon a}{\Sigma}\beta\sigma'_\rho = -\frac{\epsilon a}{\Sigma}\sigma'_\rho. \qquad (3.3.57)$$

Now by (3.3.52) we see that $U - U'$ has the same sign as $1 - \beta$. But ϵ_1 in (3.3.44) has the same sign as $U - U'$, and hence $\epsilon_1(1 - \beta)$ is positive. Since Σ and σ'_1 are positive, we get from (3.3.56)

$$\Sigma = \epsilon_1(1-\beta)\sigma'_1, \qquad (3.3.58)$$

and so the focal event lies at

$$x_1 = \frac{a\epsilon_2}{\beta-1}, \quad x_2 = 0, \quad x_3 = 0, \qquad (3.3.59)$$

the value of t remaining undetermined. We have, in fact, as indeed we must expect, a line of focal events in space-time, parallel to the

t-axis, this line cutting t = const. in the position given by (3.3.59). We recall that $\epsilon_2 = +1$ or -1 according as the incident ray strikes a convex or concave surface of the sphere.

Let us now analyse the possible occurrences when parallel incident rays strike a fixed spherical potential-jump of radius a, with potential energy U' outside and U inside. The rays strike a convex surface and so $\epsilon_2 = 1$ in preceding formulae.

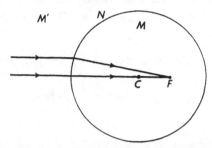

Fig. 3.12. Parallel rays incident on a fixed spherical potential-jump. Case $U < U'$ (potential hole). Focal point at F.

Suppose first that $U < U'$ (potential hole). Then every incident ray is refracted into the sphere, and as in (3.3.59) we get a focal point of the second class, as shown in fig. 3.12, at a distance

$$CF = \frac{a}{\beta - 1} \qquad (3.3.60)$$

beyond the centre of the sphere; here β is, by (3.3.53),

$$\beta = \frac{1}{m\gamma'v'}\left[\left(\frac{U'-U}{c} + m\gamma'c\right)^2 - m^2c^2\right]^{\frac{1}{2}} = \frac{(\gamma^2-1)^{\frac{1}{2}}}{(\gamma'^2-1)^{\frac{1}{2}}}. \qquad (3.3.61)$$

We note that β is the ratio of refractive indices as in (3.3.26), and so we recognize in (3.3.60) a familiar formula of optics. We note that if $v' \to c$, then $\gamma' \to \infty$ and $\beta \to 1$; the focal point moves off to infinity on the right, and becomes only a virtual focus as soon as the rays meet outside the sphere; in this limit the rays pass undisturbed through the potential-jump. If we let $v' \to 0$, then $\gamma' \to 1$ and $\beta \to \infty$; the focal point moves to the centre of the sphere.

Suppose now that $U > U'$. If v' is so small as to satisfy

$$m\gamma'c^2 + U' < mc^2 + U, \qquad (3.3.62)$$

then every ray is reflected (fig. 3.13). But if v' is so large as to satisfy

$$m\gamma'c^2 + U' > mc^2 + U, \qquad (3.3.63)$$

then refraction takes place for rays making sufficiently small angles with the normal, and there is a focal point as in (3.3.60). Since

$$m\gamma'c^2 + U' = m\gamma c^2 + U, \qquad (3.3.64)$$

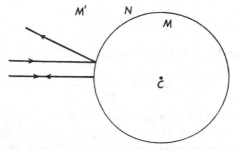

Fig. 3.13. Parallel rays incident on a fixed spherical potential-jump. Case $U > U'$, $m\gamma'c^2 + U' < mc^2 + U$. All rays reflected.

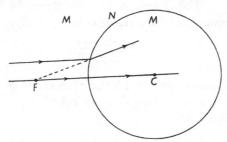

Fig. 3.14. Parallel rays incident on a fixed spherical potential-jump. Case $U > U'$, $m\gamma'c^2 + U' > mc^2 + U$. Focal point at F.

as in (3.3.12), we have $\gamma < \gamma'$, and so by (3.3.61) $\beta < 1$. Thus the focal point F lies in front of the centre of the sphere (fig. 3.14); it is of course a virtual focus. If we let $v' \to c$, then $\gamma' \to \infty$ and $\beta \to 1$, so that the focal point moves off to infinity on the left; in this limit the rays are undisturbed by the potential-jump. We cannot let $v' \to 0$, because that would violate (3.3.63); this inequality defines a lower bound for v' (say v_0') given by

$$m\gamma_0'c^2 + U' = mc^2 + U, \qquad (3.3.65)$$

so that

$$\frac{v_0'}{c} = \frac{(U - U')^{\frac{1}{2}}(U - U' + 2mc^2)^{\frac{1}{2}}}{U - U' + mc^2}. \qquad (3.3.66)$$

If $v' = v_0'$, then (by comparison of (3.3.64) and (3.3.65)) $\gamma = 1$, $v = 0$, and so the particles are reduced to rest after refraction. The corresponding β is zero, and so the limiting focal point is at the point of incidence of the central ray.

Should we wish to speak of waves instead of rays, we note that by (2.5.11) the incident wave velocity is

$$u'_\rho = -\mathrm{i}c\frac{\sigma'_\rho\,\sigma'_4}{\sigma'_\pi\,\sigma'_\pi}. \qquad (3.3.67)$$

By (3.3.11) and (3.3.24) this becomes

$$u'_\rho = v'_\rho\frac{m\gamma'c^2 + U'}{m\gamma'v'^2}. \qquad (3.3.68)$$

Here the question of gauge transformation (§2.3) becomes important. For if potential energy is regarded as undefined to within an additive constant, then the numerator in (3.3.68) can take any value. By (3.3.68) the wave and ray velocities have the same line in space, but they will be *opposed* in direction if $m\gamma'c^2 + U'$ is negative.

For the magnitude of the wave velocity, (3.3.68) gives

$$u' = \frac{|m\gamma'c^2 + U'|}{m\gamma'v'}. \qquad (3.3.69)$$

From this we can obtain a quadratic equation to give v' in terms of u'; but the ambiguous sign in the solution must be fixed by special considerations.

3.4. Rays and waves for a charged particle in a given electromagnetic field

Let us accept the usual equations of motion for a charged particle in a given electromagnetic field:

$$m\frac{\mathrm{d}^2 x_r}{\mathrm{d}s^2} = \frac{e}{c^2}F_{rm}\frac{\mathrm{d}x_m}{\mathrm{d}s}. \qquad (3.4.1)$$

Here m is the proper mass, e the charge, and F_{rs} the skew-symmetric tensor specifying the field, connected with the electric vector E_ρ and the magnetic vector H_ρ by

$$\left.\begin{array}{lll} E_1 = \mathrm{i}F_{14}, & E_2 = \mathrm{i}F_{24}, & E_3 = \mathrm{i}F_{34}, \\ H_1 = F_{23}, & H_2 = F_{31}, & H_3 = F_{12}. \end{array}\right\} \qquad (3.4.2)$$

Assuming that F_{rs} satisfies one half of Maxwell's equations,

$$F_{rs,t} + F_{st,r} + F_{tr,s} = 0 \qquad (3.4.3)$$

(the comma denoting partial differentiation), then there exists a potential 4-vector A_r such that

$$\frac{e}{c} F_{rs} = A_{s,r} - A_{r,s}. \qquad (3.4.4)$$

It is convenient for notational purposes to insert the factor e/c here. Note that the other half of Maxwell's equations are not required.

Then the equations of motion (3.4.1) may be written

$$mc \frac{\mathrm{d}^2 x_r}{\mathrm{d}s^2} = (A_{m,r} - A_{r,m}) \frac{\mathrm{d}x_m}{\mathrm{d}s}, \qquad (3.4.5)$$

which, as one can verify at once, are the Euler-Lagrange equations (2.1.4) corresponding to the variational equation

$$\delta \int f(x, \alpha)\, \mathrm{d}s = 0, \quad f(x, \alpha) = mc(-\alpha_r \alpha_r)^{\frac{1}{2}} - A_r \alpha_r, \qquad (3.4.6)$$

where $\alpha_r = \mathrm{d}x_r/\mathrm{d}s$. Thus the motion of a charged particle in a given electromagnetic field presents itself as a particular case of the general theory of Chapter II with the medium-function (3.4.6.), and so the whole subject of electron optics is covered by this general theory in a completely relativistic way, i.e. without any approximations based on the smallness of v/c.

Since the 4-potential A_r is not determined by the field F_{rs}, but may be subjected to the gauge transformation

$$A_r^* = A_r + \frac{\partial \phi}{\partial x_r} \qquad (3.4.7)$$

(where ϕ is arbitrary), we recognize that the waves derived from (3.4.6) are not gauge-invariant (cf. §2.3). Since it is not desirable to tie the theory down to any particular normalization of A_r, we shall proceed on the assumption that A_r is any 4-vector consistent with (3.4.4).

The basic formulae are given by substituting in the formulae of Chapter II the particular expression (3.4.6) for $f(x, \alpha)$. The slowness 4-vector is, by (2.1.9),

$$\sigma_r = -\frac{\partial f}{\partial \alpha_r} = mc\alpha_r + A_r. \qquad (3.4.8)$$

Introducing the particle velocity v_ρ, we have

$$\alpha_\rho = \gamma v_\rho/c, \quad \alpha_4 = i\gamma, \quad \gamma = (1 - v^2/c^2)^{-\frac{1}{2}}, \qquad (3.4.9)$$

and so the slowness 4-vector is related to the particle velocity by

$$\sigma_\rho = m\gamma v_\rho + A_\rho, \quad \sigma_4 = im\gamma c + A_4. \qquad (3.4.10)$$

Elimination of α_r from (3.4.8) by means of $\alpha_r \alpha_r = -1$ gives the slowness equation (2.1.14) in the form

$$\Omega(\sigma, x) = (\sigma_r - A_r)(\sigma_r - A_r) + m^2 c^2 = 0. \qquad (3.4.11)$$

The Hamilton-Jacobi equation is

$$\left(\frac{\partial V}{\partial x_r} + A_r\right)\left(\frac{\partial V}{\partial x_r} + A_r\right) + m^2 c^2 = 0. \qquad (3.4.12)$$

We note that the 4-vectors α_r and σ_r do not have the same direction; they are related geometrically as shown in fig. 3.15.

By (2.1.19) the unit 3-wave has the equation

$$S: \quad mc(-y_r y_r)^{\frac{1}{2}} - A_r y_r = 1. \quad (3.4.13)$$

It is a quadric 3-space with equation

$$(m^2 c^2 \delta_{rs} + A_r A_s) y_r y_s + 2 A_r y_r + 1 = 0. \qquad (3.4.14)$$

The 3-surface of slowness is, by (2.1.20),

$$W: \quad (y_r - A_r)(y_r - A_r) + m^2 c^2 = 0. \quad (3.4.15)$$

It is a pseudosphere with centre at A_r and radius mc. We recall that S and W are polar reciprocals with respect to the unit pseudosphere.

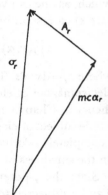

Fig. 3.15. Space-time diagram of the relationship between the 4-vectors σ_r, $mc\alpha_r$ and A_r in electromagnetic field.

Equation (2.5.11) gives us the following expression for the wave velocity u_ρ in terms of the particle velocity v_ρ:

$$u_\rho = -ic\frac{\sigma_\rho \sigma_4}{\sigma_\pi \sigma_\pi} = -ic\frac{(m\gamma v_\rho + A_\rho)(im\gamma c + A_4)}{(m\gamma v_\pi + A_\pi)(m\gamma v_\pi + A_\pi)}. \quad (3.4.16)$$

The direction of u_ρ is thus in general different from the direction of v_ρ.

But in one rather important case these vectors have the same direction, viz. the case of an electrostatic field specified by

$$A_\rho = 0, \quad A_4 = \frac{ie}{c} U(x_1, x_2, x_3), \qquad (3.4.17)$$

U being the electostatic potential. By (3.4.2) and (3.4.4) the field is

$$E_\rho = -U_{,\rho}, \quad H_\rho = 0. \tag{3.4.18}$$

By (3.4.16) the wave velocity is

$$u_\rho = \frac{v_\rho}{m\gamma v^2}(m\gamma c^2 + eU), \tag{3.4.19}$$

so that u_ρ and v_ρ have the same direction, as stated above; further

$$um\gamma v = m\gamma c^2 + eU, \tag{3.4.20}$$

in which we recognize the left-hand side as the product of wave velocity and momentum, and the right-hand side as energy.

For a magnetostatic field we take $A_4 = 0$ and A_ρ independent of x_4. In this case (3.4.16) simplifies only slightly, by the omission of A_4. The vectors u_ρ and v_ρ have different directions.

In the general static case, for which A_r is independent of x_4, the Euler-Lagrange equations of motion (2.1.4) possess the first integral

$$\sigma_4 = -\frac{\partial f}{\partial \alpha_4} = mc\alpha_4 + A_4 = iK/c, \tag{3.4.21}$$

where K is a real constant for the ray. Thus in the electrostatic case, we have the integral of energy

$$m\gamma c^2 + eU = K, \tag{3.4.22}$$

and in the magnetostatic case

$$m\gamma c^2 = K. \tag{3.4.23}$$

Note that these results are consequences of the absence of x_4 from $f(x, \alpha)$; it is in fact an ignorable coordinate. If x_1 were ignorable instead of x_4, we would have a different first integral, viz. $\sigma_1 = \text{const.}$

The statical case lends itself particularly well to treatment by Hamilton's dynamical method as described at the end of §2.1. It may be said that this method has acquired its popularity largely through the accident that t is so often absent from physical problems, in the sense that it does not occur in $f(x, \alpha)$; it was for these cases that Hamilton designed his dynamical method, although of course it works even if t is not ignorable. However, in a problem in which x_1 was ignorable instead of t, it would be better to define the Lagrangian, not as in (2.1.34), but by

$$f(x_2, x_3, x_4, \alpha_1, \alpha_2, \alpha_3, \alpha_4)\,\mathrm{d}s = L(x_2, x_3, x_4, \dot{x}_2, \dot{x}_3, \dot{x}_4)\,\mathrm{d}x_1, \tag{3.4.24}$$

in which the dot means $\mathrm{d}/\mathrm{d}x_1$.

The refractive index in a statical electromagnetic field. However, at (2.10.44) we had another way of looking at statical cases, a way more in accord with our general approach. This was further pursued in the work at (3.3.32), where we found the refractive index for a particle in a statical potential field. Let us now be more general and ask: what is the refractive index g of (2.10.44) for a particle in a statical electromagnetic field?

Note that *statical* here means that A_r is independent of x_4 (i.e. independent of t). This implies that the tensor F_{rs} is also independent of t, since F_{rs} is obtained from A_r by differentiation. However, if we are given F_{rs} independent of t we can surely find an appropriate A_r independent of t connected with F_{rs} by (3.4.4), and so we may regard the independence of t as applying to both F_{rs} and A_r.

To find the refractive index g, we have then as in (2.10.45)

$$\left.\begin{aligned}(\sigma_\rho - A_\rho)(\sigma_\rho - A_\rho) + (\sigma_4 - A_4)^2 + m^2 c^2 = 0, \\ \beta_\rho = \theta(\sigma_\rho - A_\rho), \quad g = \theta\sigma_\rho(\sigma_\rho - A_\rho).\end{aligned}\right\} \quad (3.4.25)$$

Here β_ρ are the direction cosines of the ray in space $(\beta_\rho \beta_\rho = 1)$ Then

$$\left.\begin{aligned}\theta &= [(\sigma_\rho - A_\rho)(\sigma_\rho - A_\rho)]^{-\frac{1}{2}} = [-(\sigma_4 - A_4)^2 - m^2 c^2]^{-\frac{1}{2}}, \\ \sigma_\rho &= A_\rho + \beta_\rho \theta^{-1}, \\ g &= \theta\beta_\rho\theta^{-1}(A_\rho + \beta_\rho\theta^{-1}) = \beta_\rho A_\rho + \theta^{-1},\end{aligned}\right\}$$

$$(3.4.26)$$

and so the refractive index is

$$g(x_1, x_2, x_3, \beta_1, \beta_2, \beta_3) = \beta_\rho A_\rho + (\beta_\rho\beta_\rho)^{\frac{1}{2}}[-(\sigma_4 - A_4)^2 - m^2 c^2]^{\frac{1}{2}},$$

$$(3.4.27)$$

the factor $(\beta_\rho\beta_\rho)^{\frac{1}{2}}$ being inserted to make g positive homogeneous of degree unity in β_ρ, as required in Hamiltonian optics. *This means that the rays in space (i.e. the tracks of charged particles), in the case when A_r is independent of t, satisfy the variational principle*

$$\delta\int\{[-(\sigma_4 - A_4)^2 - m^2 c^2]^{\frac{1}{2}}\,\mathrm{d}s_0 + A_\rho\,\mathrm{d}x_\rho\} = 0, \quad (3.4.28)$$

the end-points being fixed under the variation and σ_4 being an assigned and unvaried constant; $\mathrm{d}s_0$ is the line element of space $(\mathrm{d}s_0^2 = \mathrm{d}x_\rho\,\mathrm{d}x_\rho)$.

If we use (3.4.10) and write A_4 as in (3.4.17) (but with A_ρ now not vanishing), then the variational principle reads

$$\delta\int\{mc(\gamma^2 - 1)^{\frac{1}{2}}\,\mathrm{d}s_0 + A_\rho\,\mathrm{d}x_\rho\} = 0, \quad (3.4.29)$$

where γ is given by $m\gamma c^2 + eU = \text{const.}$ (3.4.30)

(Cf. (3.3.36) for the case $A_\rho = 0$; there we had U instead of eU to represent potential energy.)

Wave velocity and ray (or particle) velocity in a general electromagnetic field. Let us now return to the case of a general electromagnetic field and complete the discussion of the connexion between the wave velocity u_ρ and the ray velocity v_ρ, ray velocity being of course only an alternative expression for particle velocity. We have already expressed u_ρ in terms of v_ρ in (3.4.16). It is well to draw the reader's attention to the fact that that formula and those which follow are obtained *without any reference to wave-length or frequency, neither of which concepts appear in the theory of de Broglie waves as developed in this book before Chapter* IV.

As was pointed out earlier, the whole question of the connexion between wave velocity u_ρ and particle velocity v_ρ is contained in the seven equations (2.5.19); in the present case of a particle in an electromagnetic field, these read

$$\left.\begin{aligned} u_\rho &= -\mathrm{i}c\,\frac{\sigma_\rho\sigma_4}{\sigma_\pi\sigma_\pi}, \quad v_\rho = \mathrm{i}c\,\frac{\sigma_\rho - A_\rho}{\sigma_4 - A_4}, \\ \Omega(\sigma, x) &= (\sigma_\rho - A_\rho)(\sigma_\rho - A_\rho) + (\sigma_4 - A_4)^2 + m^2c^2 = 0. \end{aligned}\right\} \quad (3.4.31)$$

Elimination of σ_r from these equations leads to (3.4.16), expressing u_ρ in terms of v_ρ. But should we wish to obtain v_ρ in terms of u_ρ, we may proceed as follows.

The first of (3.4.31) gives

$$u^2 = -c^2\,\frac{\sigma_4^2}{\sigma_\pi\sigma_\pi}, \quad \frac{\sigma_\pi\sigma_\pi}{\sigma_4} = -\frac{c^2}{u^2}\sigma_4, \quad (3.4.32)$$

and hence $$\sigma_\rho = -\frac{1}{\mathrm{i}c}u_\rho\frac{\sigma_\pi\sigma_\pi}{\sigma_4} = -\mathrm{i}c\frac{u_\rho}{u^2}\sigma_4. \quad (3.4.33)$$

Substituting this in the last of (3.4.31), we get

$$\left(\mathrm{i}c\sigma_4\frac{u_\rho}{u^2} + A_\rho\right)\left(\mathrm{i}c\sigma_4\frac{u_\rho}{u^2} + A_\rho\right) + (\sigma_4 - A_4)^2 + m^2c^2 = 0, \quad (3.4.34)$$

or, expressed as a quadratic equation for σ_4,

$$\sigma_4^2\left(1 - \frac{c^2}{u^2}\right) + 2\sigma_4\left(\mathrm{i}cA_\rho\frac{u_\rho}{u^2} - A_4\right) + A_r A_r + m^2c^2 = 0. \quad (3.4.35)$$

The roots are

$$\sigma_4 = \frac{1}{1-c^2/u^2}\left\{-icA_\rho\frac{u_\rho}{u^2}+A_4\right.$$

$$\left.\pm\left[\left(icA_\rho\frac{u_\rho}{u^2}-A_4\right)^2-\left(1-\frac{c^2}{u^2}\right)(A_rA_r+m^2c^2)\right]^{\frac{1}{2}}\right\}.$$

$$(3.4.36)$$

Apparently the only criterion to determine the sign here is the fact that $(\sigma_4-A_4)/i=m\gamma c>0$, but this may not suffice. Leaving this ambiguity unresolved, we see that the second of (3.4.31) with (3.4.33) gives the following expression for particle velocity v_ρ in terms of wave velocity u_ρ:

$$v_\rho = -ic\frac{icu_\rho\sigma_4/u^2+A_\rho}{\sigma_4-A_4}. \qquad (3.4.37)$$

In this expression σ_4 is to be replaced by the expression (3.4.36).

3.5. De Broglie waves from a source-event in a uniform electrostatic field

Let us now investigate the de Broglie waves from a source-event in a uniform electrostatic field, which we shall define by the 4-potential
$$A_\rho=0, \quad A_4=-ipmcx_3, \qquad (3.5.1)$$
where p is a constant. The field is given by (3.4.2) and (3.4.4); all components vanish except

$$E_3=iF_{34}=\frac{ic}{e}A_{4,3}=\frac{pmc^2}{e}. \qquad (3.5.2)$$

We shall take p positive, which amounts only to a proper choice of the direction of the x_3-axis.

The equations of motion (3.4.1) may be written

$$\frac{d\alpha_r}{ds}=\frac{e}{mc^2}F_{rm}\alpha_m, \qquad (3.5.3)$$

and so we get

$$\frac{d\alpha_1}{ds}=0, \quad \frac{d\alpha_2}{ds}=0, \quad \frac{d\alpha_3}{ds}=-ip\alpha_4, \quad \frac{d\alpha_4}{ds}=ip\alpha_3. \qquad (3.5.4)$$

Integration gives
$$\left.\begin{aligned}
\alpha_1&=A, \quad \alpha_2=B,\\
\alpha_3&=C\cosh ps+D\sinh ps,\\
\alpha_4&=iC\sinh ps+iD\cosh ps.
\end{aligned}\right\} \qquad (3.5.5)$$

These constants of integration are not independent; on account of the identity $\alpha_r \alpha_r = -1$ and the fact that $\alpha_4/i > 0$ (since α_r points into the future), we have

$$A^2 + B^2 + C^2 - D^2 = -1, \quad D > 0. \tag{3.5.6}$$

Let us take $x'_r = 0$ as the source-event and measure s from it along all the rays. Then integration of (3.5.5) gives

$$\left.\begin{aligned} y_1 &= 2Aq, \quad y_2 = 2Bq, \\ y_3 &= C\sinh 2q + D(\cosh 2q - 1), \\ y_4 &= C(\cosh 2q - 1) + D\sinh 2q, \end{aligned}\right\} \tag{3.5.7}$$

in terms of dimensionless real coordinates y_r and a dimensionless parameter q, defined by

$$y_\rho = px_\rho, \quad y_4 = px_4/i = pct, \quad q = \tfrac{1}{2}ps. \tag{3.5.8}$$

We can find the value of q (or equivalently s) corresponding to any event y_r, reached by the rays by solving (3.5.7) for A, B, C, D, in terms of q and substituting the result in (3.5.6). The solution of (3.5.7) is

$$\left.\begin{aligned} A &= \tfrac{1}{2}y_1/q, \quad B = \tfrac{1}{2}y_2/q, \\ C &= \tfrac{1}{2}(y_3\cosh q - y_4\sinh q)/\sinh q, \\ D &= \tfrac{1}{2}(y_4\cosh q - y_3\sinh q)/\sinh q. \end{aligned}\right\} \tag{3.5.9}$$

Then substitution in (3.5.6) gives

$$\frac{y_1^2 + y_2^2}{q^2} + \frac{y_3^2 - y_4^2}{\sinh^2 q} + 4 = 0. \tag{3.5.10}$$

To obtain q in terms of y_r, we would have to solve this equation.

The characteristic function is

$$\begin{aligned} V &= \int_{P'}^{P} (mc + ipmcx_3\alpha_4)\,\mathrm{d}s \\ &= mcs + ipmc\int_{P'}^{P} x_3\,\mathrm{d}x_4 \\ &= \frac{2mc}{p}\left(q - \frac{1}{2}\int_{P'}^{P} y_3\,\mathrm{d}y_4\right) \\ &= \frac{2mc}{p}\left\{q - \int_0^q [C\sinh 2q + D(\cosh 2q - 1)]\right. \\ &\qquad\qquad\qquad \left.\times [C\sinh 2q + D\cosh 2q]\,\mathrm{d}q\right\}. \end{aligned}$$

$$\tag{3.5.11}$$

This gives

$$\frac{pV}{2mc} = q[1 + \tfrac{1}{2}(C^2 - D^2)] - \tfrac{1}{8}(C^2 + D^2)\sinh 4q$$
$$+ \tfrac{1}{2}D^2 \sinh 2q - CD \cosh 2q \sinh^2 q. \quad (3.5.12)$$

In this we substitute for C and D from (3.5.9) and obtain

$$\frac{pV}{2mc} = q + \tfrac{1}{8}\left(\coth q - \frac{q}{\sinh^2 q}\right)(y_4^2 - y_3^2) - \tfrac{1}{4}y_3 y_4. \quad (3.5.13)$$

If we eliminate q from (3.5.10) and (3.5.13), we get a relation of the form
$$F(y_1, y_2, y_3, y_4, V) = 0. \quad (3.5.14)$$

The de Broglie 3-waves from a given source-event for a charged particle in a uniform electrostatic field are given by putting $V = \text{const.}$ in this equation. To get the de Broglie 2-waves at time t, we solve (3.5.13) for y_3, obtaining, say,

$$y_3 = G(y_4, V, q). \quad (3.5.15)$$

By (3.5.10) we have then

$$y_1^2 + y_2^2 = q^2\left[\frac{y_4^2 - G^2}{\sinh^2 q} - 4\right]. \quad (3.5.16)$$

On putting $V = \text{const.}$ and $y_4 (= pct) = \text{const.}$ in (3.5.15) and (3.5.16), *we have in parametric form (the parameter being q) the equation of a de Broglie 2-wave at time t.* It is of course a surface of revolution.

In this way the 2-waves can be plotted. But we can go further explicitly if we confine our attention to events on the x_3-axis. For then, by (3.5.10),

$$\left.\begin{aligned}
y_4^2 - y_3^2 = 4\sinh^2 q, \quad 2\sinh q = (y_4^2 - y_3^2)^{\frac{1}{2}}, \\
q = \log \tfrac{1}{2}[(y_4^2 - y_3^2)^{\frac{1}{2}} + (y_4^2 - y_3^2 + 4)^{\frac{1}{2}}], \\
\coth q = (y_4^2 - y_3^2 + 4)^{\frac{1}{2}}/(y_4^2 - y_3^2)^{\frac{1}{2}},
\end{aligned}\right\} \quad (3.5.17)$$

and so (3.5.13) reads

$$\frac{pV}{2mc} = \tfrac{1}{2}\log \tfrac{1}{2}[(y_4^2 - y_3^2)^{\frac{1}{2}} + (y_4^2 - y_3^2 + 4)^{\frac{1}{2}}]$$
$$+ \tfrac{1}{8}(y_4^2 - y_3^2)^{\frac{1}{2}}(y_4^2 - y_3^2 + 4)^{\frac{1}{2}} - \tfrac{1}{4}y_3 y_4. \quad (3.5.18)$$

This gives the history of the intersections of the de Broglie 2-waves with the x_3-axis.

If $y_4^2 - y_3^2$ is large, we may write (3.5.18) in the approximate form

$$\frac{pV}{2mc} = \tfrac{1}{8}(y_4^2 - y_3^2 - 2y_3 y_4). \qquad (3.5.19)$$

The two places where the 3-wave $V = $ const. cuts the x_3-axis at time t ($y_4 = pct$) are thus given by

$$y_3 = y_4\left[-1 \pm \left(2 - \frac{4pV}{mcy_4^2} \right)^{\frac{1}{2}} \right]. \qquad (3.5.20)$$

The velocity u of the wave along the x_3-axis is obtained by differentiating (3.5.18) with V fixed; if we use the approximation (3.5.19) instead of (3.5.18), we get

$$u = c\frac{dy_3}{dy_4} = c\frac{y_4 - y_3}{y_4 + y_3} = c\frac{ct - x_3}{ct + x_3}, \qquad (3.5.21)$$

this formula being an approximation, good for $p^2(c^2t^2 - x_3^2)$ large. Thus if we take up a station at position x_3 and wait for a long time, the waves are ultimately passing with velocity c in that direction in which the particle would move from rest under the action of the field. It is interesting that a reversal of the sign of x_3 inverts the fraction in (3.5.21). Thus, if u and u' are the wave velocities at the same time t at points $x_3 = \pm a$, then

$$uu' = c^2. \qquad (3.5.22)$$

3.6. De Broglie waves for a particle in a central field of force

Let us take again the medium-function as in (3.3.1):

$$f(x, \alpha) = mc(-\alpha_r \alpha_r)^{\frac{1}{2}} - iU\alpha_4/c, \quad U = U(x_1, x_2, x_3). \qquad (3.6.1)$$

In (3.3.5) we recognized U as potential energy. In the theory which follows, we take U to be a function of r only, where $r^2 = x_\pi x_\pi$. We shall later make the special choice

$$U = ee'/r, \qquad (3.6.2)$$

and then our problem will be that of the motion of a particle of proper mass m and charge e in the electrostatic field of a charge e', fixed at the origin. But for the present we shall work with a general function $U(r)$.

For the slowness 4-vector we have

$$\sigma_\pi = mc\alpha_\pi, \quad \sigma_4 = mc\alpha_4 + iU/c, \qquad (3.6.3)$$

and hence, eliminating α_r by $\alpha_r \alpha_r = -1$, the slowness equation

$$2\Omega(\sigma, x) = \sigma_\pi \sigma_\pi + (\sigma_4 - iU/c)^2 + m^2 c^2 = 0. \qquad (3.6.4)$$

The Hamiltonian equations of the rays are then, as in (2.1.17),

$$\left.\begin{array}{ll} \dfrac{\mathrm{d}x_\rho}{\mathrm{d}w} = \dfrac{\partial\Omega}{\partial\sigma_\rho} = \sigma_\rho, & \dfrac{\mathrm{d}x_4}{\mathrm{d}w} = \dfrac{\partial\Omega}{\partial\sigma_4} = \sigma_4 - iU/c, \\[3mm] \dfrac{\mathrm{d}\sigma_\rho}{\mathrm{d}w} = -\dfrac{\partial\Omega}{\partial x_\rho} = \dfrac{ix_\rho}{cr}\dfrac{\mathrm{d}U}{\mathrm{d}r}\left(\sigma_4 - \dfrac{iU}{c}\right), & \dfrac{\mathrm{d}\sigma_4}{\mathrm{d}w} = -\dfrac{\partial\Omega}{\partial x_4} = 0. \end{array}\right\} \quad (3.6.5)$$

These yield the integral of energy

$$\sigma_4 = imcA, \qquad (3.6.6)$$

where A is a dimensionless constant, and the integral of angular momentum

$$x_\rho \sigma_\pi - x_\pi \sigma_\rho = mcC_{\rho\pi}, \qquad (3.6.7)$$

where $C_{\rho\pi}$ ($= -C_{\pi\rho}$) are constants with the dimensions of a length.

Consider now the set of rays in space-time for which A and $C_{\rho\pi}$ have assigned values. We must note that a set of rays, selected arbitrarily, does not necessarily form a system of rays with 3-waves associated with it. In order that it may form a system in this sense, it is necessary (and sufficient) that there exists some 3-space Σ_0 such that at each event on Σ_0 the normal has the direction of the slowness 4-vector σ_r obtained from the ray by the formula $\sigma_r = -\partial f/\partial\alpha_r$ (cf. §2.2). By (2.2.6) this is equivalent to the condition that

$$\sigma_r \mathrm{d}x_r = -\mathrm{d}V, \qquad (3.6.8)$$

an exact differential.

We shall now show that the set of rays with assigned A and $C_{\rho\pi}$ is in fact a system of rays in this sense, but with a dimensional degeneracy. We simplify the algebra by making a special choice of axes in space. The skew-symmetric matrix $C_{\rho\pi}$ defines a direction in space, and if we take that direction for x_3-axis, we get

$$C_{23} = C_{31} = 0, \quad C_{12} \neq 0. \qquad (3.6.9)$$

From these and (3.6.7) we obtain

$$x_3 C_{12} = 0, \quad \sigma_3 C_{12} = 0, \qquad (3.6.10)$$

and hence $x_3 = 0, \qquad \sigma_3 = 0, \qquad (3.6.11)$

thus formally deriving the otherwise well-known fact that the orbits lie in a single plane.

The expression which we have to prove to be an exact differential now reads

$$\sigma_r \, dx_r = \sigma_1 \, dx_1 + \sigma_2 \, dx_2 + \sigma_4 \, dx_4. \tag{3.6.12}$$

Collecting together (3.6.4), (3.6.6), (3.6.7), (3.6.11), we have

$$\left.\begin{aligned}
\sigma_1^2 + \sigma_2^2 &= m^2 c^2 \left[\left(A - \frac{U}{mc^2} \right)^2 - 1 \right], \\
\sigma_4 &= imcA, \\
x_1 \sigma_2 - x_2 \sigma_1 &= mcC_{12}.
\end{aligned}\right\} \tag{3.6.13}$$

Squaring the last of these, we have

$$r^2(\sigma_1^2 + \sigma_2^2) - (x_1 \sigma_1 + x_2 \sigma_2)^2 = m^2 c^2 C_{12}^2, \tag{3.6.14}$$

and so

$$(x_1 \sigma_1 + x_2 \sigma_2)^2 = r^2 m^2 c^2 \left[\left(A - \frac{U}{mc^2} \right)^2 - 1 \right] - m^2 c^2 C_{12}^2. \tag{3.6.15}$$

Let us define

$$F(r) = \left[\left(A - \frac{U}{mc^2} \right)^2 - 1 - \frac{C_{12}^2}{r^2} \right]^{\frac{1}{2}}; \tag{3.6.16}$$

then

$$x_1 \sigma_1 + x_2 \sigma_2 = \epsilon m c r \, F(r), \quad \epsilon = \pm 1. \tag{3.6.17}$$

Combining this with the last of (3.6.13), we solve for σ_1 and σ_2:

$$\left.\begin{aligned}
\sigma_1 &= \epsilon m c \frac{x_1}{r} F(r) - mc \frac{x_2}{r^2} C_{12}, \\
\sigma_2 &= \epsilon m c \frac{x_2}{r} F(r) + mc \frac{x_1}{r^2} C_{12}.
\end{aligned}\right\} \tag{3.6.18}$$

By (3.6.11) we have $r^2 = x_1^2 + x_2^2$, and so

$$\begin{aligned}
\sigma_1 \, dx_1 + \sigma_2 \, dx_2 &= \epsilon m c \, F(r) \, dr + \frac{mc}{r^2} C_{12}(x_1 \, dx_2 - x_2 \, dx_1) \\
&= \epsilon m c \, F(r) \, dr + mc C_{12} \, d\phi, \tag{3.6.19}
\end{aligned}$$

where ϕ is the polar angle such that

$$x_1 = r \cos \phi, \quad x_2 = r \sin \phi.$$

Thus the form (3.6.12) is an exact differential:

$$-dV = \sigma_r \, dx_r = \epsilon m c \, F(r) \, dr + mc C_{12} \, d\phi - mc^2 A \, dt. \tag{3.6.20}$$

We recognize then that the selected set of rays (degenerate since contained in $x_3 = 0$) has associated with it a set of waves given by

$$V(r, \phi, t) = \text{const.}, \quad x_3 = 0, \tag{3.6.21}$$

where

$$-V = \epsilon m c \int F(r) \, dr + mc C_{12} \phi - mc^2 A t + K(\epsilon) \quad (\epsilon = \pm 1), \tag{3.6.22}$$

$K(\epsilon)$ being arbitrary constants. The equations (3.6.21) define a set of 2-spaces in space-time, and we may speak of these as *de Broglie 2-waves*. But we are to remember that these are a degenerate form of the de Broglie 3-waves which occur in the general case. Each 2-wave is the history of a 1-*wave*, obtained by putting t = const. Thus a 1-wave has the equations

$$\phi = -\frac{\epsilon}{C_{12}}\int F(r)\,\mathrm{d}r + (mc^2At - V - K(\epsilon))/(mcC_{12}),$$

$$x_3 = 0, \quad t = \text{const.}, \quad (3.6.23)$$

V being a constant. From this equation we can draw the pattern of 1-waves in the plane $x_3 = 0$. As t increases this pattern rotates rigidly about the origin with angular velocity

$$\omega = cA/C_{12}. \tag{3.6.24}$$

Since $F(r)$ must be real, the rays and waves are confined to that portion of the plane $x_3 = 0$ for which

$$\left(A - \frac{U}{mc^2}\right)^2 - 1 - \frac{C_{12}^2}{r^2} \geqslant 0. \tag{3.6.25}$$

The constants A and C_{12} may be such that this inequality is satisfied for all values of r. But let us consider the *oscillatory case* for which this inequality is satisfied only for $r_1 \leqslant r \leqslant r_2$, the expression in (3.6.25) vanishing at the ends of this range, so that

$$F(r_1) = 0, \quad F(r_2) = 0. \tag{3.6.26}$$

The rays then touch the two circles $r = r_1$ and $r = r_2$. Let us make the expressions (3.6.22) and (3.6.23) more precise by writing a definite integral:

$$-V = \epsilon mc\int_{r_1}^r F(r)\,\mathrm{d}r + mcC_{12}\phi - mc^2At + K(\epsilon), \quad (3.6.27)$$

$$\phi = -\frac{\epsilon}{C_{12}}\int_{r_1}^r F(r)\,\mathrm{d}r + L(\epsilon), \tag{3.6.28}$$

where $\epsilon = \pm 1$ and

$$L(\epsilon) = (mc^2At - V - K(\epsilon))/(mcC_{12}). \tag{3.6.29}$$

If we proceed along a 1-wave as given by (3.6.28) (V and t, and hence L, being constants), we have

$$\frac{\mathrm{d}\phi}{\mathrm{d}r} = -\frac{\epsilon}{C_{12}}F(r); \tag{3.6.30}$$

hence, by (3.6.26), $\mathrm{d}\phi/\mathrm{d}r=0$ for $r=r_1$ and for $r=r_2$. *Thus the de Broglie* I*-waves meet the bounding circles* $r=r_1$ *and* $r=r_2$ *at right angles.*

Equation (3.6.28) gives two I-waves, one for $\epsilon = +\mathrm{I}$ and the other for $\epsilon = -\mathrm{I}$. These are of the forms PP' and QQ'' as shown in fig. 3.16, which is drawn for positive C_{12}, the sense of ϕ increasing being counter-clockwise. These curves are symmetric in the sense that

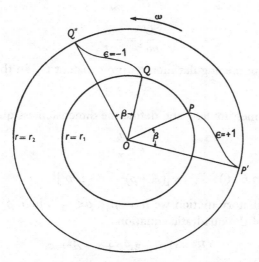

Fig. 3.16. Space picture of de Broglie I-waves PP', QQ'' for oscillatory motion in a central field of force.

QQ'' may be obtained from PP' by first rotating the latter rigidly about O until P coincides with Q and then reflecting in the line of the radius OQ. We have

$$\phi(P)=L(\mathrm{I}), \qquad \phi(P')=-\frac{\mathrm{I}}{C_{12}}\int_{r_1}^{r_2}F(r)\,\mathrm{d}r+L(\mathrm{I}), \left.\begin{array}{l}\\[2em]\end{array}\right\} \quad (3.6.31)$$
$$\phi(Q)=L(-\mathrm{I}), \quad \phi(Q'')=\frac{\mathrm{I}}{C_{12}}\int_{r_1}^{r_2}F(r)\,\mathrm{d}r+L(-\mathrm{I}).$$

The angle $P'OP$ is equal to QOQ'' and the common value is

$$\beta=\frac{\mathrm{I}}{C_{12}}\int_{r_1}^{r_2}F(r)\,\mathrm{d}r. \qquad (3.6.32)$$

The whole pattern of de Broglie 1-waves at time t (for arbitrary values of V) is obtained from the two curves of fig. 3.16 by rotating rigidly about O through all angles. As t changes, this pattern rotates with the angular velocity given by (3.6.24).

The case of a Coulomb field. Let us now turn to the particular case of a Coulomb field, with U as in (3.6.2). We shall take ee' negative and introduce the length

$$k = -\frac{ee'}{mc^2}, \tag{3.6.33}$$

so that

$$\frac{U}{mc^2} = -\frac{k}{r}. \tag{3.6.34}$$

Let us write the angular momentum constant C_{12} in the form

$$C_{12} = kB, \tag{3.6.35}$$

B being dimensionless, and define the dimensionless quantity ρ by

$$\rho = \frac{k}{r}. \tag{3.6.36}$$

Then by (3.6.16) $F(r) = [(A+\rho)^2 - 1 - B^2\rho^2]^{\frac{1}{2}}. \tag{3.6.37}$

For oscillatory motion we have $\rho_2 \leqslant \rho \leqslant \rho_1$, where ρ_1 and ρ_2 are the roots of the quadratic equation

$$(B^2 - 1)\rho^2 - 2A\rho + 1 - A^2 = 0. \tag{3.6.38}$$

There are two basic constants, and we may use either the pair (A, B) or the pair (ρ_1, ρ_2); they are connected by the relations

$$\rho_1 + \rho_2 = \frac{2A}{B^2 - 1}, \quad \rho_1\rho_2 = \frac{1 - A^2}{B^2 - 1}. \tag{3.6.39}$$

Since ρ_1, ρ_2 are to be positive and $F(r)$ real for $\rho_2 \leqslant \rho \leqslant \rho_1$, the constants A and B satisfy the inequalities

$$A > 0, \quad A^2 < 1, \quad B^2 > 1, \quad A^2B^2 - B^2 + 1 \geqslant 0. \tag{3.6.40}$$

To express one pair of constants in terms of the other pair, we have

$$\left.\begin{aligned}
\rho_1 &= \frac{1}{B^2 - 1}[A + (A^2B^2 - B^2 + 1)^{\frac{1}{2}}], \\
\rho_2 &= \frac{1}{B^2 - 1}[A - (A^2B^2 - B^2 + 1)^{\frac{1}{2}}],
\end{aligned}\right\} \tag{3.6.41}$$

and
$$A = \frac{1}{\rho_1 + \rho_2} \left[-\rho_1\rho_2 + \{\rho_1^2\rho_2^2 + (\rho_1 + \rho_2)^2\}^{\frac{1}{2}} \right],$$
$$B^2 = \frac{1}{(\rho_1 + \rho_2)^2} \left[\rho_1^2 + \rho_2^2 + 2\{\rho_1^2\rho_2^2 + (\rho_1 + \rho_2)^2\}^{\frac{1}{2}} \right]. \qquad (3.6.42)$$

We may write (3.6.37) in the form

$$F(r) = (B^2 - 1)^{\frac{1}{2}} \{(\rho_1 - \rho)(\rho - \rho_2)\}^{\frac{1}{2}}, \qquad (3.6.43)$$

and the integral which occurs in (3.6.27) and (3.6.28) is

$$\int_{r_1}^{r} F(r)\,dr = k(B^2 - 1)^{\frac{1}{2}} \int_{\rho}^{\rho_1} \{(\rho_1 - \rho)(\rho - \rho_2)\}^{\frac{1}{2}} \frac{d\rho}{\rho^2}$$

$$= k(B^2 - 1)^{\frac{1}{2}} \left[\frac{1}{\rho} \{(\rho_1 - \rho)(\rho - \rho_2)\}^{\frac{1}{2}} \right.$$

$$\left. + \frac{\rho_1 + \rho_2}{(\rho_1\rho_2)^{\frac{1}{2}}} (\tfrac{1}{2}\pi - \psi) - \pi + 2\theta \right], \quad (3.6.44)$$

where θ and ψ are acute angles defined by

$$\tan\theta = \left(\frac{\rho - \rho_2}{\rho_1 - \rho}\right)^{\frac{1}{2}}, \quad \tan\psi = \left(\frac{\rho_1}{\rho_2}\right)^{\frac{1}{2}} \left(\frac{\rho - \rho_2}{\rho_1 - \rho}\right)^{\frac{1}{2}}. \qquad (3.6.45)$$

With the above value of the integral, the characteristic function V is given by

$$-\frac{V}{mc} = \epsilon \int_{r_1}^{r} F(r)\,dr + kB\phi - cAt + \frac{K(\epsilon)}{mc} \quad (\epsilon = \pm 1), \quad (3.6.46)$$

and the 1-waves may be plotted from the equation

$$\phi = -\frac{\epsilon}{B} (B^2 - 1)^{\frac{1}{2}} \left[\frac{1}{\rho} \{(\rho_1 - \rho)(\rho - \rho_2)\}^{\frac{1}{2}} \right.$$

$$\left. + \frac{\rho_1 + \rho_2}{(\rho_1\rho_2)^{\frac{1}{2}}} (\tfrac{1}{2}\pi - \psi) - \pi + 2\theta \right] + L(\epsilon). \qquad (3.6.47)$$

We note that

$$(B^2 - 1)^{-\frac{1}{2}} \int_{r_1}^{r_2} F(r)\,dr = k \int_{\rho_2}^{\rho_1} \{(\rho_1 - \rho)(\rho - \rho_2)\}^{\frac{1}{2}} \frac{d\rho}{\rho^2}$$

$$= \tfrac{1}{2}\pi k \left(\frac{\rho_1 + \rho_2}{(\rho_1\rho_2)^{\frac{1}{2}}} - 2\right). \qquad (3.6.48)$$

The angle β of (3.6.32) is

$$\beta = \frac{1}{kB} \int_{r_1}^{r_2} F(r)\, dr$$

$$= \frac{1}{2} \frac{\pi}{B} (B^2 - 1)^{\frac{1}{2}} \left(\frac{\rho_1 + \rho_2}{(\rho_1 \rho_2)^{\frac{1}{2}}} - 2 \right)$$

$$= \frac{\pi}{B} (B^2 - 1)^{\frac{1}{2}} \left[\frac{A}{\{(1 - A^2)(B^2 - 1)\}^{\frac{1}{2}}} - 1 \right]. \qquad (3.6.49)$$

These formulae will be used in the quantization of the hydrogenic atom in Chapter IV. It should be noted that so far we have proceeded without introducing wave-length or frequency; the de Broglie 1-waves form a continuum of moving curves in the plane of the rays.

PRIMITIVE QUANTIZATION

4.1. Quantization for plane waves for a free particle

So far the theory has been confined to the domain of geometrical mechanics. The de Broglie waves have had no phase, no frequency, no wave-length. It is by *quantization* that these things are introduced, in much the same way as they are introduced in the transition from geometrical optics to physical optics.

Rays and waves for a free particle were discussed in §3.1, and in particular the rays and waves corresponding to an initial plane 3-wave Σ_0. Then, as in fig. 3.2, we have a system of parallel rays in space-time ($\alpha_r = $ const.) and a system of parallel plane 3-waves, the rays being orthogonal to the 3-waves. If we choose Σ_0 to pass through the origin, then, as in (3.1.12), the characteristic function is

$$V(P) = -\sigma_r x_r, \qquad (4.1.1)$$

where σ_r is constant, and the de Broglie 3-waves have the equations

$$\sigma_r x_r = \text{const.} \qquad (4.1.2)$$

To carry out the quantization, we introduce the scalar wave equation

$$\psi_{,rr} - \frac{m^2 c^2}{\hbar^2} \psi = 0, \qquad (4.1.3)$$

where $\hbar = h/2\pi$, h being Planck's constant, and

$$\psi_{,rr} = \Box \psi = \Delta \psi - \frac{1}{c^2} \frac{\partial^2 \psi}{\partial t^2}.$$

We ask: are there solutions ψ of (4.1.3) such that $\psi = $ const. on each of the de Broglie 3-waves (4.1.2)? To answer this question, we put

$$\psi = F(\sigma_r x_r). \qquad (4.1.4)$$

Then, σ_r being constant,

$$\psi_{,r} = \sigma_r F', \quad \psi_{,rr} = \sigma_r \sigma_r F''. \qquad (4.1.5)$$

But, by (3.1.6) we have $\sigma_r \sigma_r = -m^2 c^2$, $\qquad (4.1.6)$

and so (4.1.4) satisfies (4.1.3) provided F satisfies

$$F'' + \frac{1}{\hbar^2}F = 0. \tag{4.1.7}$$

The required solution is therefore

$$\psi = \psi_0 \cos(\sigma_r x_r/\hbar + \epsilon) = \psi_0 \cos(V/\hbar - \epsilon), \tag{4.1.8}$$

where ψ_0 and ϵ are constants.

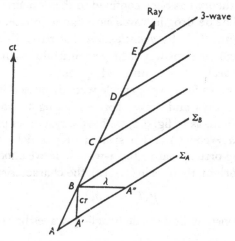

Fig. 4.1. Space-time diagram of quantization of plane de Broglie 3-waves for a free particle.

$$V_E - V_D = V_D - V_C = V_C - V_B = V_B - V_A = h,$$
$$AB = BC = CD = DE = h/mc = \text{absolute de Broglie wave-length},$$
$$\nu = 1/\tau = \text{relative de Broglie frequency} = m\gamma c^2/h,$$
$$\lambda = \text{relative de Broglie wave-length} = h/m\gamma v.$$

An important property of this solution is to be noted. If we pass along a ray from any event A to the next event B with the same phase, the increment in V is such that

$$(V_B - V_A)/\hbar = 2\pi, \quad V_B - V_A = \int_A^B f \, ds = mcAB = h, \tag{4.1.9}$$

where f is the medium-function and AB is the Minkowskian interval. *Thus the action along a ray between events of the same phase is precisely Planck's constant h.*

The events of equal phase, $A, B, C, D, E, \dots$ (fig. 4.1) laid off on a ray in space-time, are spaced at Minkowskian intervals h/mc. Thus h/mc might be called the *absolute de Broglie wave-length*.

But in discussing de Broglie waves one usually considers the *relative* wave-length λ and the *relative* frequency ν ($\nu = 1/\tau$, where τ is the *relative* period). These quantities are of course not Lorentz-invariant. In the present simple case of plane 3-waves, they can be found accurately in the familiar forms, and we shall carry out the work here in detail in order to make the theory complete. It should be noted that the derivation involves the simple character of the waves and could be carried through only in an approximate form (for short wave-length) if we were dealing with curved waves.

To find τ we draw the 3-waves Σ_A, Σ_B through consecutive events A, B of equal phase (fig. 4.1). We draw through B a straight line parallel to the t-axis, meeting Σ_A at A', say. Then $c\tau = A'B$ (the Minkowskian separation). Now in an obvious notation

$$AB_r = AA'_r + A'B_r. \qquad (4.1.10)$$

Since the rays are orthogonal to the 3-waves, we have $AB_r AA'_r = 0$, and since $A'B_r$ is parallel to the t-axis, only its fourth component survives. On multiplying (4.1.10) by AB_r, we get therefore

$$-AB^2 = AB_r AB_r = AB_r A'B_r = AB_4 A'B_4 = AB_4 ic\tau. \qquad (4.1.11)$$

But $$AB_4 = AB\alpha_4 = ABi\gamma, \qquad (4.1.12)$$

where $\gamma^{-2} = 1 - v^2/c^2$, v being the particle velocity. Then, by (4.1.9) and (4.1.11),

$$c\tau\gamma = AB = (V_B - V_A)/(mc) = h/mc. \qquad (4.1.13)$$

Thus the de Broglie relative period and frequency are

$$\tau = \frac{h}{mc^2\gamma}, \quad \nu = \frac{mc^2\gamma}{h}. \qquad (4.1.14)$$

By (3.1.11) the de Broglie 2-waves have velocity

$$u = c^2/v, \qquad (4.1.15)$$

and so the relative de Broglie wave-length is

$$\lambda = u\tau = \frac{h}{mv\gamma}. \qquad (4.1.16)$$

This relative wave-length is indicated as BA'' in fig. 4.1.

4.2. Quantization for waves from a source-event

Again we consider a free particle as in §3.1, but now, instead of a system of parallel rays, we take the system of rays emanating from a source-event P'. Taking P' at the origin, the de Broglie 3-waves are, as in (3.1.13),

$$c^2 t^2 - r^2 = (V/mc)^2 = \text{const.} \qquad (4.2.1)$$

To quantize these waves, we seek a solution of the wave equation (4.1.3) which is constant on each of the pseudospheres (4.2.1). So we put

$$\psi = F(R), \quad R^2 = -x_r x_r = c^2 t^2 - r^2 = (V/mc)^2. \qquad (4.2.2)$$

Then

$$\psi_{,r} = -\frac{x_r}{R} F', \quad \psi_{,rr} = -\frac{4}{R} F' - R \frac{\mathrm{d}}{\mathrm{d}R}\left(\frac{F'}{R}\right), \qquad (4.2.3)$$

and so the satisfaction of (4.1.3) demands

$$R \frac{\mathrm{d}}{\mathrm{d}R}\left(\frac{F'}{R}\right) + \frac{4}{R} F' + \frac{m^2 c^2}{\hbar^2} F = 0, \qquad (4.2.4)$$

or

$$F'' + \frac{3}{R} F' + \frac{m^2 c^2}{\hbar^2} F = 0. \qquad (4.2.5)$$

The domain of the rays is the interior of the future null-cone drawn from the origin, and we seek a solution without singularity in that domain. The solution is (Watson, 1944, p. 97)

$$\psi = F(R) = \psi_0 \frac{J_1(mcR/\hbar)}{mcR/\hbar}, \qquad (4.2.6)$$

where ψ_0 is an arbitrary constant and J_1 the Bessel function of the first order. This solution may also be written

$$\psi = \psi_0 \frac{J_1(V/\hbar)}{V/\hbar}. \qquad (4.2.7)$$

The de Broglie 3-waves for the rays from a source-event were shown in fig. 3.4. That diagram is still valid, but the 3-waves have now been provided with phase by virtue of (4.2.7). If we proceed along a ray, we start with a maximum of ψ at P'; ψ oscillates, becoming zero at the zeros of J_1, that is, for

$$V/\hbar = 3 \cdot 8317, \quad 7 \cdot 0156, \quad 10 \cdot 1735, \quad \dots \qquad (4.2.8)$$

(Watson, 1944, p. 748). The asymptotic form for large V is

$$\psi \sim \psi_0 \left(\frac{2}{\pi}\right)^{\frac{1}{2}} \left(\frac{\hbar}{V}\right)^{\frac{3}{2}} \left[\sin\left(\frac{V}{\hbar} - \tfrac{1}{4}\pi\right) + \frac{3}{8}\frac{\hbar}{V}\sin\left(\frac{V}{\hbar} + \tfrac{1}{4}\pi\right)\right] \qquad (4.2.9)$$

(Watson, 1944, p. 195). Thus for large values of V, consecutive events A, B on a ray with equal phases are such that

$$V_B - V_A = \int_A^B f \, ds = mc \, AB = h, \qquad (4.2.10)$$

approximately. Note that this is only *approximate*, whereas (4.1.9) was *exact*. Near the source-event P', events of equal phase are slightly further separated from one another. Thus if A refers to the first of (4.2.8) and B to the third, we have

$$V_B - V_A = \int_A^B f \, ds = mc \, AB = 6 \cdot 3418 \hbar = 1 \cdot 0093 h. \quad (4.2.11)$$

Since $1 \cdot 0093$ differs from unity by less than 1% (and if we used higher zeros of J_1 we would get a number still closer to unity), we may say that *for de Broglie 3-waves from a source-event the Minkowskian separation between adjacent events of equal phase on a ray is approximately h/mc, and approaches this value as a limit as we pass out along the ray.* This separation was *accurately h/mc* in the case of plane waves (§ 4.1). These considerations suggest the primitive quantization based on an absolute action quantum h, proposed in the next section.

On account of the curvature of the 3-waves from a source-event, the relative de Broglie frequency and wave-length no longer admit accurate expressions as in (4.1.14) and (4.1.16). Approximate expressions can be found, but these will be covered by the discussion in the next section.

Since the source-event appears to be a rather fundamental concept, it is interesting to construct some kinematical pictures of the corresponding de Broglie 2-waves.

We note that by (4.2.1) we have $V = 0$ on the null-cone, and hence by (4.2.7) ψ has the constant value ψ_0 over the null-cone. We may call the null-cone the *leading 3-wave*; the corresponding *leading 2-wave* is a spherical wave advancing with velocity c, its radius being ct.

For purposes of graphical representation, the deviation from unity of the factor in (4.2.11) is not significant, and we may in fair approximation obtain events of equal phase by measuring off the absolute de Broglie wave-length h/mc along the rays in space-time.

This means (cf. (4.2.1)) that we get the set of 3-waves with the same phase as the leading wave by writing

$$c^2t^2 - r^2 = (nh/mc)^2 \quad (n=0, 1, 2, \ldots). \qquad (4.2.12)$$

Thus the set of 2-waves at time t are to be plotted from the formula

$$r^2 = c^2t^2 - (nh/mc)^2 \quad (n=0, 1, 2, \ldots); \qquad (4.2.13)$$

the velocities of these 2-waves are given by

$$u = \frac{dr}{dt} = \frac{c^2t}{r} = \frac{c}{\left(1 - \dfrac{n^2h^2}{m^2c^4t^2}\right)^{\frac{1}{2}}} \quad (n=0, 1, 2, \ldots). \qquad (4.2.14)$$

Here $n=0$ corresponds to the leading wave.

It is not easy to visualize the rather complicated kinematics of these 2-waves; figs. 4.2–4.7 illustrate some representative cases.

In fig. 4.2, where $t=h/mc^2$, the leading wave $n=0$ has a radius h/mc and velocity c; the wave $n=1$ is just starting, with infinite velocity. In fig. 4.3, where $t=5h/mc^2$, the radii and velocities of the waves are as follows:

Number of wave [n in (4.2.13), (4.2.14)]	Radius $= \dfrac{h}{mc} \times$	Velocity $= c \times$
0	5	1
1	4·899	1·021
2	4·583	1·091
3	4	1·25
4	3	1·667
5	0	∞

For $t = 10^{20}h/mc^2$ the region of disturbance is a sphere of radius $10^{20}h/mc$ (fig. 4.4). To get an idea of the pattern of the 2-waves of equal phase, we pick out three sample regions at P' (the centre), Q (half-way out) and R (on the boundary). The patterns in these neighbourhoods are shown in figs. 4.5, 4.6, 4.7.

Near P' we have (fig. 4.5)

Number of wave	Radius $= \dfrac{h}{mc} \times$	Velocity $= c \times$
$10^{20} - 3$	$2 \cdot 449 \times 10^{10}$	$4 \cdot 082 \times 10^9$
$10^{20} - 2$	$2 \quad \times 10^{10}$	$5 \quad \times 10^9$
$10^{20} - 1$	$1 \cdot 414 \times 10^{10}$	$7 \cdot 071 \times 10^9$
10^{20}	0	∞

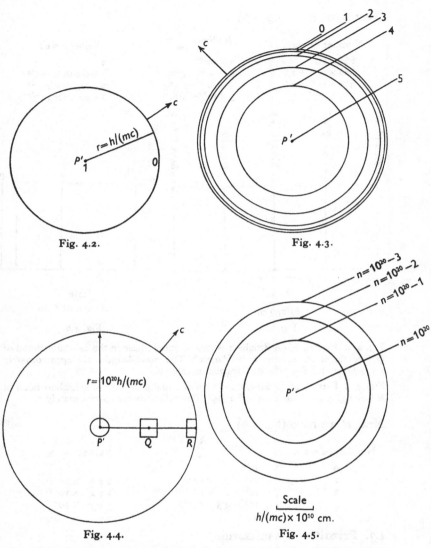

Fig. 4.2.

Fig. 4.3.

Fig. 4.4.

Scale
$h/(mc) \times 10^{10}$ cm.

Fig. 4.5.

Fig. 4.2. De Broglie 2-waves of equal phase from a source-event $P'(x')$ observed at time $t = h/mc^2$. The leading wave $n = 0$ is advancing with speed c: the next wave $n = 1$ is starting from the point P' with infinite velocity.

Fig. 4.3. De Broglie 2-waves of equal phase from a source-event $P'(x')$ observed at time $t = 5h/mc^2$.

Fig. 4.4. Region of disturbance caused by de Broglie waves from a source-event $P'(x')$ observed at time $t = 10^{20} h/mc^2$. For the wave-patterns at P', Q and R, see the following figures.

Fig. 4.5. Pattern of de Broglie 2-waves of equal phase in the neighbourhood of the source-event P' (see fig. 4.4), at time $t = 10^{20} h/mc^2$.

Near Q we have (fig. 4.6)

Number of wave	Radius $= \dfrac{h}{mc} \times$	Velocity $= c \times$
866×10^{17}	5×10^{19}	2
$866 \times 10^{17} + 1$	$5 \times 10^{19} - 1 \cdot 732$	$2 + 6 \cdot 928 \times 10^{-20}$
$866 \times 10^{17} + 2$	$5 \times 10^{19} - 3 \cdot 464$	$2 + 1 \cdot 386 \times 10^{-19}$
$866 \times 10^{17} + 3$	$5 \times 10^{19} - 5 \cdot 196$	$2 + 2 \cdot 078 \times 10^{-19}$

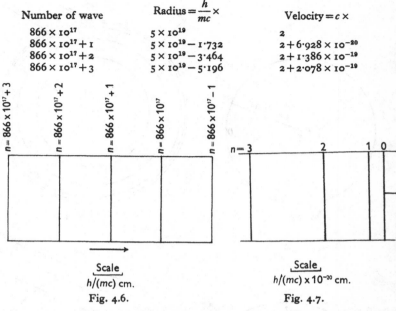

Scale
$h/(mc)$ cm.
Fig. 4.6.

Scale
$h/(mc) \times 10^{-20}$ cm.
Fig. 4.7.

Fig. 4.6. Pattern of de Broglie 2-waves of equal phase in the neighbourhood of Q (see fig. 4.4), at time $t = 10^{20} \, h/mc^2$. The wave-lengths are approximately $1 \cdot 732 \, h/mc$ and the velocities approximately $2c$.

Fig. 4.7. Pattern of de Broglie 2-waves of equal phase in the neighbourhood of R (see fig. 4.4), at time $t = 10^{20} \, h/mc^2$. The velocities are approximately c.

Near R we have (fig. 4.7)

Number of wave	Radius $= \dfrac{h}{mc} \times$	Velocity $= c \times$
0	10^{20}	1
1	$10^{20} - 5 \times 10^{-21}$	$1 + 5 \times 10^{-41}$
2	$10^{20} - 2 \times 10^{-20}$	$1 + 2 \times 10^{-40}$
3	$10^{20} - 4 \cdot 5 \times 10^{-20}$	$1 + 4 \cdot 5 \times 10^{-40}$

4.3. Primitive quantization

Suppose we are given a medium-function $f(x, \alpha)$. Taking some source-event or some initial 3-wave, we obtain a system of rays with de Broglie 3-waves associated with them. This system will have a characteristic function $V(x)$.

We have quantized such systems in §§ 4.1 and 4.2. The cases there considered were very simple, because the medium-function had the simple form $f(x, \alpha) = mc(-\alpha_r \alpha_r)^{\frac{1}{2}}$, since we were dealing with a free particle, and the systems of rays selected were very simple. Thus, we did not consider the case where there is an initial 3-wave which is an arbitrary 3-space. The question now before us is this: how are we to quantize in the general case described in the preceding paragraph?

It is not proposed to lay down here any general prescription for quantization based on a wave equation, but rather to define clearly a process which we shall call *primitive quantization*. This definition is suggested partly by the results of §§ 4.1 and 4.2 and partly by the way in which, in physical optics, one makes calculations about simple interference patterns.

Primitive quantization consists in assigning throughout the space-time domain occupied by the waves a *phase factor*

$$\cos(V/\hbar + \eta), \qquad (4.3.1)$$

where η is a constant. *This means that adjacent events A and B of equal phase on a ray satisfy*

$$V_B - V_A = \int_A^B f \, ds = 2\pi\hbar = h. \qquad (4.3.2)$$

This agrees with the formula (4.1.9) for plane waves, but with one difference: we cannot now write $\int_A^B f \, ds = mc\, AB$ because we are no longer dealing with a free particle and f is not constant along a ray. If we compare (4.3.2) with the formulae (4.2.10) and (4.2.11) for the case of a free particle with given initial source-event, we note a disagreement of the order of 1 % in the neighbourhood of the source-event but agreement at great (Minkowskian) distances from it.

It would seem therefore that from a physical point of view (4.3.2) should be regarded as a *pis aller*, to be used (as similar rules are used in physical optics) whenever exact quantization by means of a wave equation cannot be performed, perhaps because we do not know what wave equation to use or perhaps because we want to reach results quickly. But possibly this primitive quantization has greater physical validity than that, for we shall find in the next section that it gives the usual fine-structure formula for the hydrogenic atom. Whatever its physical importance, it is a good thing to make the theory of primitive quantization mathematically clear.

Note that in (4.3.2) it is a definite amount of *action* that separates events of equal phase. The Minkowskian element ds is contained under the sign of integration, and it can be brought out only when we use approximations based on the smallness of h. The rest of the present section is devoted to such approximations.

If h is small, then (4.3.2) gives for the Minkowskian separation AB between adjacent events of equal phase, if we use (2.1.10),

$$AB = \frac{h}{f} = -\frac{h}{\sigma_r \alpha_r}. \qquad (4.3.3)$$

This may be regarded as the *absolute de Broglie wave-length*. To get the relative de Broglie frequency and wave-length, we proceed as in fig. 4.1, in which the elements are now to be regarded as infinitesimal on account of the smallness of h. Thus, although AB and AA' are now curves, their curvatures do not matter because the triangle is infinitesimal. We have then, as in (4.1.10),

$$AB_r = AA'_r + A'B_r. \qquad (4.3.4)$$

The reasoning following (4.1.10) cannot be used now, because in the present general case the rays are not orthogonal to the 3-waves. Instead we write, by (4.3.3),

$$AB_r = -\frac{h}{\sigma_n \alpha_n} \alpha_r. \qquad (4.3.5)$$

We multiply (4.3.4) by σ_r and remember that σ_r is orthogonal to the 3-wave, so that $AA'_r \sigma_r = 0$: thus

$$-h = A'B_r \sigma_r. \qquad (4.3.6)$$

Only the fourth component of $A'B_r$ survives and it is $A'B_4 = ic\tau$, where τ is the relative de Broglie period. Hence

$$\tau = \frac{ih}{c\sigma_4}, \quad \nu = -\frac{ic\sigma_4}{h}. \qquad (4.3.7)$$

These are the general approximate expressions (for h small) for the relative de Broglie period and frequency.

To get the relative de Broglie wave-length we refer to (2.5.19);

for wave propagation in the direction of the unit 3-vector n_ρ we have

$$\sigma_\rho = n_\rho (\sigma_\pi \sigma_\pi)^{\frac{1}{2}},\qquad (4.3.8)$$

and so

$$\lambda = u_\rho n_\rho \tau = -ic\frac{n_\rho \sigma_\rho}{\sigma_\pi \sigma_\pi}\sigma_4\frac{ih}{c\sigma_4} = \frac{h}{(\sigma_\pi \sigma_\pi)^{\frac{1}{2}}}.\qquad (4.3.9)$$

This is the approximate expression (for small h) of the relative de Broglie wave-length.

4.4. Primitive quantization for a central field of force and for the hydrogenic atom

In § 3.6 we discussed rays and waves for a particle in a central field of force, specializing at (3.6.33) to the case of a Coulomb field, that is, to a hydrogenic atom treated as a Kepler problem. We shall now apply the process of primitive quantization to such systems, treating first the case of a general central field and then specializing for the hydrogenic atom.

Fig. 3.16 shows de Broglie 1-waves travelling round in the annular region bounded by the circles $r = r_1$ and $r = r_2$. The characteristic function V is as in (3.6.27):

$$-V = \epsilon mc\int_{r_1}^{r} F(r)\,dr + mcC_{12}\phi - mc^2At + K(\epsilon)\quad (\epsilon = \pm 1),\quad (4.4.1)$$

r, ϕ being polar coordinates in the plane of the rays. To approach the problem of quantization, we consider as in fig. 4.8 two 1-waves, PP' and QQ', both corresponding to $\epsilon = +1$ and both drawn for the same value of t. On each of them V is a constant (by the definition of a wave), and by (4.4.1) we have

$$-V_Q + V_P = -V_{Q'} + V_{P'} = mcC_{12}(\phi_Q - \phi_P) = mcC_{12}(\phi_{Q'} - \phi_{P'}).\qquad (4.4.2)$$

The wave QQ' is obtained from PP' by a rigid body rotation about the centre through an angle $\phi_Q - \phi_P$.

Introducing the phase factor (4.3.1), the condition that PP' and QQ' shall be adjacent waves of equal phase is

$$mcC_{12}(\phi_Q - \phi_P) = \pm h.\qquad (4.4.3)$$

The same phase must be restored when we increase ϕ by 2π, and so we get as *first quantum condition*

$$2\pi mcC_{12} = jh\quad (j = \pm 1,\ \pm 2,\ \pm 3,\ ...).\qquad (4.4.4)$$

The same condition applies to the waves with $\epsilon = -1$. We may also write this first quantum condition in the geometrical form

$$\phi_Q - \phi_P = 2\pi/j \quad (j = \pm 1, \pm 2, \pm 3, \ldots). \tag{4.4.5}$$

A second quantum condition is obtained by comparing the waves for $\epsilon = +1$ with those for $\epsilon = -1$. On the inner circle $r = r_1$ we have, by (4.4.1),
$$-V = mcC_{12}\phi - mc^2At + K(\epsilon), \tag{4.4.6}$$

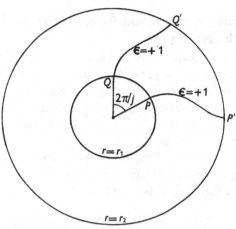

Fig. 4.8. Space picture of two de Broglie 1-waves of equal phase for a central field (not drawn to scale).
$$-V_Q + V_P = -V_{Q'} + V_{P'} = mcC_{12}(\phi_Q - \phi_P) = h.$$

and on the outer circle $r = r_2$

$$-V = \epsilon mc \int_{r_1}^{r_2} F(r)\,dr + mcC_{12}\phi - mc^2At + K(\epsilon). \tag{4.4.7}$$

These formulae give two distributions of phase on each circle, one distribution for $\epsilon = +1$ and the other distribution for $\epsilon = -1$. *We now make the demand that there shall be a unique distribution of phase on each circle.* This demand, for $r = r_1$, requires that $K(1) - K(-1)$ shall be a multiple of h; then, for $r = r_2$, this demand gives us the *second quantum condition*

$$2mc \int_{r_1}^{r_2} F(r)\,dr = 2\pi n\hbar = nh \quad (n = 0, 1, 2, \ldots). \tag{4.4.8}$$

The value $n = 0$ corresponds to $r_1 = r_2$, that is, to the case of a circular orbit—a further degeneracy.

This second quantum condition may be interpreted geometrically as follows. In fig. 4.9 P is any point on the inner circle $r = r_1$. PP' and PP'' are 1-waves through P, with $\epsilon = +1$ for PP' and $\epsilon = -1$ for PP''. Then since $V_P = V_{P'} = V_{P''}$, we have by (4.4.6) and (4.4.7)

$$\left. \begin{aligned} mc \int_{r_1}^{r_2} F(r)\, dr + mcC_{12}(\phi_{P'} - \phi_P) &= 0, \\ -mc \int_{r_1}^{r_2} F(r)\, dr + mcC_{12}(\phi_{P''} - \phi_P) &= 0. \end{aligned} \right\} \qquad (4.4.9)$$

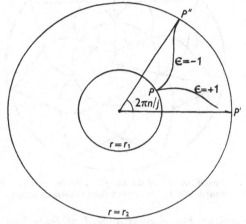

Fig. 4.9. De Broglie 1-waves for $\epsilon = +1$ and $\epsilon = -1$
(not drawn to scale).

But by symmetry $\phi_{P'} - \phi_P = \phi_P - \phi_{P''}$, and so

$$mcC_{12}(\phi_{P'} - \phi_{P''}) = 2mc \int_{r_1}^{r_2} F(r)\, dr. \qquad (4.4.10)$$

Hence, by (4.4.8), $\qquad \phi_{P'} - \phi_{P''} = \dfrac{nh}{mcC_{12}}, \qquad (4.4.11)$

and so, by (4.4.3), $\qquad \phi_{P'} - \phi_{P''} = n(\phi_Q - \phi_P), \qquad (4.4.12)$

Q referring to the next wave of the same phase, as in fig. 4.8.

The geometrical meaning of the quantum numbers j and n may be described as follows. There are j points of equal phase on the inner circle $r = r_1$ and j points of equal phase on the outer circle also, the angle subtended at the centre by two adjacent points of equal phase being $2\pi/j$. As for n, if we draw the 1-waves for $\epsilon = +1$ and for $\epsilon = -1$ from any point on the inner circle, these waves cut the outer circle at two points which subtend at the centre an angle $2\pi n/j$, i.e. n times the angle $2\pi/j$ mentioned above, or twice the angle β of fig. 3.16, viz. $\beta = n\pi/j$.   (4.4.13)

The type of pattern formed by the 1-waves of equal phase may best be understood by looking at fig. 4.10, which is drawn for the values $j = 7$, $n = 2$.

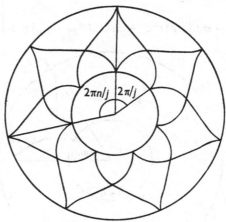

Fig. 4.10. Space picture of de Broglie 1-waves of equal phase for a central field; $j = 7$, $n = 2$ (not drawn to scale).

To summarize: *For a central field of force (not necessarily a Coulomb field) the geometrical quantum conditions (see figs. 4.8, 4.9, 4.10) are* $\phi_Q - \phi_P = 2\pi/j$, $\phi_{P'} - \phi_{P'} = n(\phi_Q - \phi_P)$,   (4.4.14)
and the equivalent analytical conditions are

$$mcC_{12} = j\hbar \quad (j = \pm 1, \pm 2, \pm 3, \ldots),$$
$$mc \int_{r_1}^{r_2} F(r)\,\mathrm{d}r = \pi n\hbar \quad (n = 0, 1, 2, \ldots), \qquad (4.4.15)$$

where C_{12} is the constant of angular momentum, as in (3.6.7), *and $F(r)$ is as in* (3.6.16).

The hydrogenic atom. We turn now to the hydrogenic atom. By (3.6.35) and (3.6.48) the quantum conditions (4.4.15) become

$$mckB = j\hbar \quad (j = \pm 1, \pm 2, \pm 3, \ldots),$$
$$mck(B^2 - 1)^{\frac{1}{2}} \left[\frac{A}{(1 - A^2)^{\frac{1}{2}} (B^2 - 1)^{\frac{1}{2}}} - 1 \right] = n\hbar \quad (n = 0, 1, 2, \ldots).$$
$$(4.4.16)$$

These are two equations, to be solved for the dimensionless numbers A and B. It is convenient to introduce the fine-structure constant and associated notation:

$$\alpha = \frac{e^2}{\hbar c}, \quad e' = -Ze, \quad \frac{mck}{\hbar} = -\frac{mcee'}{\hbar mc^2} = Z\alpha. \quad (4.4.17)$$

Then
$$B = \frac{j}{Z\alpha} \quad (j = \pm 1, \pm 2, \pm 3, \ldots), \quad (4.4.18)$$

and so, by (3.6.7) and (3.6.35), the quantization of angular momentum is (since $\sigma_\rho = mc\alpha_\rho = m\gamma v_\rho$)

$$m\gamma(x_1 v_2 - x_2 v_1) = mcC_{12} = mckB = \frac{j\hbar}{Z\alpha} \quad (j = \pm 1, \pm 2, \pm 3, \ldots).$$
$$(4.4.19)$$

As for energy, we have to solve (4.4.16) for A: we obtain

$$A = \left[1 + \frac{Z^2\alpha^2}{\{n + (j^2 - Z^2\alpha^2)^{\frac{1}{2}}\}^2} \right]^{-\frac{1}{2}}$$
$$(j = \pm 1, \pm 2, \pm 3, \ldots; n = 0, 1, 2, \ldots). \quad (4.4.20)$$

By (3.6.13) and (3.6.34)

$$\left. \begin{array}{l} imcA = \sigma_4 = mc\alpha_4 + iU/c = im\gamma c + iU/c, \\ mc^2 A = m\gamma c^2 + U = mc^2 + E, \\ E = mc^2(\gamma - 1) + U, \\ U = ee'/r, \end{array} \right\} \quad (4.4.21)$$

where U is the potential energy and E the total energy (kinetic + potential, and so by (4.4.20) the quantized energies are

$$E = mc^2(A - 1) = mc^2 \left\{ \left[1 + \frac{Z^2\alpha^2}{\{n + (j^2 - Z^2\alpha^2)^{\frac{1}{2}}\}} \right]^{-\frac{1}{2}} - 1 \right\}$$
$$(j = \pm 1, \pm 2, \pm 3, \ldots; n = 0, 1, 2, \ldots). \quad (4.4.22)$$

This is the usual fine-structure formula, as obtained from Dirac's equations (Dirac, 1930, p. 254) or by direct application of the rules of quantization (Sommerfeld, 1934, pp. 102, 256). From a formal point of view, Sommerfeld's derivation has a great similarity to the derivation given above. However, the present method seems to offer some advantages: it is part of a consistent and comprehensive theory of de Broglie waves and it offers a fuller intuitive grasp of the situation, for we can visualize the rotating pattern of the de Broglie waves and the quantum conditions as required here are very simple. This may be of general interest as creating some doubt of the assertion sometimes made that quantum theory removes physics from the domain of direct intuitive perception derived from kinematical pictures.

We can draw diagrams of the de Broglie waves to scale, and this is done in figs. 4.11–4.14. The case $n = 0$ does not deserve a diagram, since then we have a circular orbit, and all we could show on it would be j equally spaced point-waves of equal phase. The patterns for $n \neq 0$ are rather interesting.

The diagrams are drawn for cases where $Z\alpha$ is small, relative errors of the order $(Z\alpha)^2$ being neglected. Since $\alpha = \frac{1}{137}$, this means that for $Z \leqslant 13$ the errors are less that one per cent or of that order.

Expansion of (4.4.20) in powers of $Z\alpha$ gives

$$A = 1 - \frac{1}{2}\frac{(Z\alpha)^2}{(n+|j|)^2} - \frac{(Z\alpha)^4}{2(n+|j|)^4}\left(\frac{n}{|j|} + \frac{1}{4}\right) + O[(Z\alpha)^6],$$

$$1 - A^2 = \frac{(Z\alpha)^2}{(n+|j|)^2} + \frac{n}{|j|}\frac{(Z\alpha)^4}{(n+|j|)^4} + O[(Z\alpha)^6],$$

$$A^2B^2 - B^2 + 1 = \frac{n(n+2|j|)}{(n+|j|)^2} - \frac{n|j|}{(n+|j|)^4}(Z\alpha)^2 + O[(Z\alpha)^4],$$

$$\frac{1}{B^2-1} = \frac{(Z\alpha)^2}{j^2} + \frac{(Z\alpha)^4}{j^4} + O[(Z\alpha)^6],$$

$$\frac{|j|}{j}\frac{1}{B}(B^2-1)^{\frac{1}{2}} = 1 - \frac{1}{2}\frac{(Z\alpha)^2}{j^2} + O[(Z\alpha)^4].$$

$$(4.4.23)$$

We have then from (3.6.41)

$$\rho_1 = \frac{(Z\alpha)^2}{j^2}\left(1 + \frac{[n(n+2|j|)]^{\frac{1}{2}}}{n+|j|}\right) + O[(Z\alpha)^4],$$

$$\rho_2 = \frac{(Z\alpha)^2}{j^2}\left(1 - \frac{[n(n+2|j|)]^{\frac{1}{2}}}{n+|j|}\right) + O[(Z\alpha)^4],$$

$$(4.4.24)$$

and by (3.6.36) the radii of the bounding circles are approximately

$$r_1 = \frac{r_b}{Z}\frac{j^2}{1 + \dfrac{[n(n+2|j|)]^{\frac{1}{2}}}{n+|j|}},$$

$$r_2 = \frac{r_b}{Z}\frac{j^2}{1 - \dfrac{[n(n+2|j|)]^{\frac{1}{2}}}{n+|j|}},$$

$$(4.4.25)$$

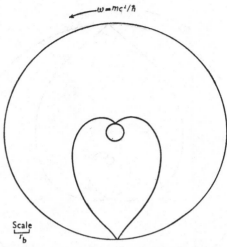

Fig. 4.11. De Broglie 1-waves of equal phase for hydrogen atom $(Z=1)$ in state $n=1, |j|=1$. Scale: $r_b = \hbar^2/me^2 =$ radius of fundamental Bohr orbit. Inner circle: $r=r_1=0.53r_b$. Outer circle: $r=r_2=7.5r_b$. For atomic number Z, reduce diagram in the ratio $1/Z$ (provided $Z\alpha$ is small): this applies also to figs. 4.12, 4.13, and 4.14.

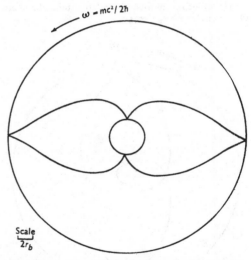

Fig. 4.12. De Broglie 1-waves of equal phase for hydrogen atom $(Z=1)$ in state $n=1, |j|=2$. Scale: $r_b =$ radius of fundamental Bohr orbit. Inner circle: $r=r_1=2.3r_b$. Outer circle: $r=r_2=15.7r_b$.

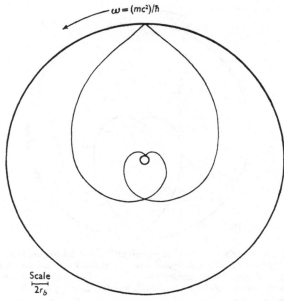

Fig. 4.13. De Broglie 1-waves of equal phase for hydrogen atom $(Z=1)$ in state $n=2$, $|j|=1$. Scale: $r_b =$ radius of fundamental Bohr orbit. Inner circle: $r=r_1=0.5r_b$. Outer circle: $r=r_2=17.5r_b$.

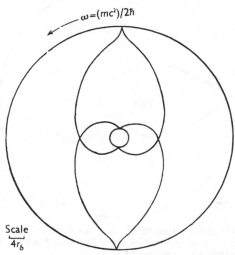

Fig. 4.14. De Broglie 1-waves of equal phase for hydrogen atom $(Z=1)$ in state $n=2$, $|j|=2$. Scale: $r_b =$ radius of fundamental Bohr orbit. Inner circle: $r=r_1=2.1r_b$. Outer circle: $r=r_2=29.8r_b$.

where r_b is the radius of the fundamental Bohr orbit,

$$r_b = \frac{1}{\alpha}\frac{\hbar}{mc} = \frac{\hbar^2}{me^2}. \qquad (4.4.26)$$

(Note that if $Z = 1$, $n = 0$, $|j| = 1$, then (4.4.25) give $r_1 = r_2 = r_b$.)

We have approximately

$$\frac{\rho_1 + \rho_2}{(\rho_1\rho_2)^{\frac{1}{2}}} = 2\left(1 + \frac{n}{|j|}\right), \qquad (4.4.27)$$

and so for purposes of plotting we take (3.6.47) in the form

$$\phi = \frac{1}{\rho}\{(\rho_1 - \rho)(\rho - \rho_2)\}^{\frac{1}{2}} + \frac{n}{|j|}\pi - 2\psi\left(1 + \frac{n}{|j|}\right) + 2\theta. \quad (4.4.28)$$

This is for the branch $\epsilon = -1$; the branch $\epsilon = +1$ is given by reflexion.

By (3.6.24) the angular velocity with which the pattern rotates is

$$\omega = \frac{cA}{C_{12}} = \frac{cA}{kB}, \qquad (4.4.29)$$

or approximately, since $A \sim 1$, $k = Z\alpha\hbar/mc$, $B = j/Z\alpha$,

$$\omega = \frac{mc^2}{j\hbar} \quad (j = \pm 1, \pm 2, \pm 3, \ldots). \qquad (4.4.30)$$

The angular velocity of an electron in the fundamental Bohr orbit is

$$\omega_b = \frac{\alpha^2 mc^2}{\hbar}, \qquad (4.4.31)$$

and so we may write (4.4.30) in the form

$$\omega = \frac{\omega_b}{j\alpha^2}. \qquad (4.4.32)$$

But ω may also be expressed in terms of a fundamental angular velocity ω_e associated with the electronic mass. Let ν_e be the fundamental frequency of the electron, defined by

$$h\nu_e = mc^2, \qquad (4.4.33)$$

and let ω_e be defined by

$$\omega_e = 2\pi\nu_e = mc^2/\hbar. \qquad (4.4.34)$$

Then (4.4.30) may be written

$$\omega = \omega_e/j \quad (j = \pm 1, \pm 2, \pm 3, \ldots). \qquad (4.4.35)$$

This simple relation is rather striking.

For plotting the de Broglie 1-waves, the following formulae, equivalent to (4.4.28), will be found useful. Here ρ and ϕ are expressed as functions of the parametric angle θ of (3.6.45), which ranges from 0 to $\frac{1}{2}\pi$. The formulae are

$$\left.\begin{array}{c}
\dfrac{\rho}{(Z\alpha)^2}=\dfrac{r_b}{rZ}=\dfrac{1}{j^2}\left(1-\dfrac{[n(n+2\,|j|)]^{\frac{1}{2}}}{n+|j|}\cos 2\theta\right), \\[4mm]
\phi=\dfrac{\sin 2\theta}{\dfrac{n+|j|}{[n(n+2\,|j|)]^{\frac{1}{2}}}-\cos 2\theta}+\left(1+\dfrac{n}{|j|}\right)(\pi-2\psi)-(\pi-2\theta), \\[6mm]
\tan\psi=\left(1+\dfrac{n}{|j|}+\dfrac{[n(n+2\,|j|)]^{\frac{1}{2}}}{|j|}\right)\tan\theta \quad (0\leqslant\psi\leqslant\tfrac{1}{2}\pi).
\end{array}\right\}$$

$$(4.4.36)$$

These formulae are obtained by using $\rho=\rho_1\sin^2\theta+\rho_2\cos^2\theta$.

It should be noted that the relationship between ϕ (the azimuthal angle) and θ (the parametric angle) involves only the *ratio* $n/|j|$. All the 1-wave forms corresponding to a given ratio $n/|j|$ are of the same *shape* but of different *sizes*, the size varying as j^2, according to the first of (4.4.36); the size varies also inversely as Z. It must, of course, be remembered that (4.4.36) are approximate formulae, valid only when $Z\alpha$ is small.

4.5. The Zeeman effect

Consider the medium-function

$$f(x,\alpha)=mc(-\alpha_r\alpha_r)^{\frac{1}{2}}-A_1\alpha_1-A_2\alpha_2-A_4\alpha_4, \qquad (4.5.1)$$

where

$$\left.\begin{array}{c}
A_1=-\dfrac{eH}{2c}x_2, \quad A_2=\dfrac{eH}{2c}x_1, \quad A_4=\dfrac{iU}{c}, \\[4mm]
\dfrac{U}{mc^2}=-\dfrac{k}{r}, \quad k=\dfrac{Ze^2}{mc^2}.
\end{array}\right\} \qquad (4.5.2)$$

This corresponds to the case of a hydrogenic atom of atomic number Z in a constant magnetic field H which points in the x_3 direction.

To discuss de Broglie waves, we proceed much as in §3.6. The components of slowness are

$$\sigma_1=mc\alpha_1+A_1, \quad \sigma_2=mc\alpha_2+A_2, \quad \sigma_3=mc\alpha_3, \quad \sigma_4=mc\alpha_4+A_4,$$

$$(4.5.3)$$

and from these equations we obtain the slowness equation

$$2\Omega(\sigma, x) = (\sigma_1 - A_1)^2 + (\sigma_2 - A_2)^2 + \sigma_3^2 + (\sigma_4 - A_4)^2 + m^2c^2 = 0.$$
$$(4.5.4)$$

The Hamiltonian equations of the rays are

$$\left.\begin{aligned}
\frac{dx_1}{dw} &= \frac{\partial\Omega}{\partial\sigma_1} = \sigma_1 - A_1, & \frac{d\sigma_1}{dw} &= -\frac{\partial\Omega}{\partial x_1} = (\sigma_2 - A_2)\frac{\partial A_2}{\partial x_1} + (\sigma_4 - A_4)\frac{\partial A_4}{\partial x_1}, \\
\frac{dx_2}{dw} &= \frac{\partial\Omega}{\partial\sigma_2} = \sigma_2 - A_2, & \frac{d\sigma_2}{dw} &= -\frac{\partial\Omega}{\partial x_2} = (\sigma_1 - A_1)\frac{\partial A_1}{\partial x_2} + (\sigma_4 - A_4)\frac{\partial A_4}{\partial x_2}, \\
\frac{dx_3}{dw} &= \frac{\partial\Omega}{\partial\sigma_3} = \sigma_3, & \frac{d\sigma_3}{dw} &= -\frac{\partial\Omega}{\partial x_3} = (\sigma_4 - A_4)\frac{\partial A_4}{\partial x_3}, \\
\frac{dx_4}{dw} &= \frac{\partial\Omega}{\partial\sigma_4} = \sigma_4 - A_4, & \frac{d\sigma_4}{dw} &= -\frac{\partial\Omega}{\partial x_4} = 0.
\end{aligned}\right\}$$
$$(4.5.5)$$

These equations possess integrals of energy and angular momentum:

$$\sigma_4 = imcA,   \tag{4.5.6}$$
$$x_1\sigma_2 - x_2\sigma_1 = mckB,   \tag{4.5.7}$$

where A and B are dimensionless constants. Of these the first is obvious and the second is easily verified.

Consider orbits lying in the plane $x_3 = 0$ with assigned values of A and B. Then $\sigma_3 = 0$ and (4.5.4) becomes

$$\sigma_1^2 + \sigma_2^2 - \frac{eH}{c}(x_1\sigma_2 - x_2\sigma_1) + \left(\frac{eH}{2c}\right)^2 r^2 - m^2c^2\left(A + \frac{k}{r}\right)^2 + m^2c^2 = 0.$$
$$(4.5.8)$$

Squaring (4.5.7), we have

$$(\sigma_1 x_1 + \sigma_2 x_2)^2 = r^2(\sigma_1^2 + \sigma_2^2) - (mckB)^2,   \tag{4.5.9}$$

and so, by (4.5.8), $\sigma_1 x_1 + \sigma_2 x_2 = \epsilon mcr\, F(r),$ (4.5.10)

where $\epsilon = \pm 1$ and

$$F(r) = \left[\left(A + \frac{k}{r}\right)^2 - 1 - \frac{k^2 B^2}{r^2} + \frac{eHkB}{mc^2} - \left(\frac{eH}{2mc^2}\right)^2 r^2\right]^{\frac{1}{2}}.   \tag{4.5.11}$$

We now solve (4.5.7), (4.5.10) for σ_1, σ_2, obtaining

$$\left.\begin{aligned}
\sigma_1 &= \epsilon mc\frac{x_1}{r}F(r) - mckB\frac{x_2}{r^2}, \\
\sigma_2 &= \epsilon mc\frac{x_2}{r}F(r) + mckB\frac{x_1}{r^2}.
\end{aligned}\right\}$$
$$(4.5.12)$$

For the characteristic function V we have then the exact differential

$$-dV = \sigma_1 dx_1 + \sigma_2 dx_2 + \sigma_4 dx_4$$
$$= \epsilon mc\, F(r)\, dr + mckB\, d\phi - mc^2 A dt, \qquad (4.5.13)$$

ϕ being the azimuthal angle. This is the same as in (3.6.20), the only difference being in the form of the function $F(r)$, which now includes two additional terms in H. If we put $\rho = k/r$, we have

$$F(r) = \left[(A+\rho)^2 - 1 - B^2\rho^2 + DB - \frac{D^2}{4\rho^2} \right]^{\frac{1}{2}}, \qquad (4.5.14)$$

where D is the dimensionless constant

$$D = \frac{eHk}{mc^2}. \qquad (4.5.15)$$

We confine our attention to oscillatory motion in which $r_1 \leqslant r \leqslant r_2$, where r_1 and r_2 are zeros of $F(r)$. If ρ_1 and ρ_2 are the corresponding values of ρ, they are solutions of

$$(A+\rho)^2 - 1 - B^2\rho^2 + DB - \frac{D^2}{4\rho^2} = 0, \qquad (4.5.16)$$

or
$$(A+\rho)^2 - 1 - \left(B\rho - \frac{D}{2\rho} \right)^2 = 0. \qquad (4.5.17)$$

So far there is no quantization. But by (4.5.13) we have

$$-V = \epsilon mc \int_{r_1}^{r} F(r)\, dr + mckB\phi - mc^2 At + K(\epsilon), \qquad (4.5.18)$$

and then the same argument as in §4.4 provides the two quantum conditions

$$2\pi mckB = jh, \qquad 2mc \int_{r_1}^{r_2} F(r)\, dr = nh, \qquad (4.5.19)$$

or, equivalently, since $mck = Z\alpha\hbar$,

$$B = \frac{j}{Z\alpha}, \qquad \frac{1}{\pi k} \int_{r_1}^{r_2} F(r)\, dr = \frac{n}{Z\alpha}$$

$$(j = \pm 1, \pm 2, \pm 3, \ldots; \; n = 0, 1, 2, \ldots). \qquad (4.5.20)$$

To compute the energy levels we would need to solve the biquadratic equation (4.5.16). This prevents us from obtaining a general exact formula, and we shall confine ourselves to the case where D is small in the sense that the term $D^2/(4\rho^2)$ in (4.5.14) and

(4.5.16) is negligible. This means that we take, instead of the accurate expression (4.5.14), the approximation

$$F(r) = (B^2 - 1)^{\frac{1}{2}} [(\rho_1 - \rho)(\rho - \rho_2)]^{\frac{1}{2}}, \qquad (4.5.21)$$

where ρ_1 and ρ_2 are the roots of the quadratic equation

$$(B^2 - 1)\rho^2 - 2A\rho + 1 - A^2 - BD = 0. \qquad (4.5.22)$$

Then, using (3.6.48), we get from (4.5.20)

$$B = \frac{j}{Z\alpha}, \quad (B^2 - 1)^{\frac{1}{2}} \left(\frac{1}{2} \frac{\rho_1 + \rho_2}{(\rho_1 \rho_2)^{\frac{1}{2}}} - 1 \right) = \frac{n}{Z\alpha}. \qquad (4.5.23)$$

Hence, by (4.5.22),

$$\frac{A}{(1 - A^2 - BD)^{\frac{1}{2}}} - (B^2 - 1)^{\frac{1}{2}} = \frac{n}{Z\alpha}, \qquad (4.5.24)$$

or

$$\frac{A}{(1 - A^2 - jD/Z\alpha)^{\frac{1}{2}}} = \frac{1}{Z\alpha} [n + (j^2 - Z^2\alpha^2)^{\frac{1}{2}}]. \qquad (4.5.25)$$

On solving for A and then using $E = mc^2(A - 1)$ as in (4.4.22), we find for the *energy levels of a hydrogenic atom in a weak magnetic field H*

$$-E = mc^2 \left[1 - \left(1 - \frac{jD}{Z\alpha} \right)^{\frac{1}{2}} \left\{ 1 + \frac{Z^2\alpha^2}{\{n + (j^2 - Z^2\alpha^2)^{\frac{1}{2}}\}^2} \right\}^{-\frac{1}{2}} \right]$$

$$(j = \pm 1, \pm 2, \pm 3, \ldots; \; n = 0, 1, 2, \ldots) \quad (4.5.26)$$

$$D = \frac{Ze^3 H}{m^2 c^4}, \quad \frac{D}{Z\alpha} = \frac{e^3 Hc\hbar}{m^2 c^4 e^2} = \frac{eH\hbar}{m^2 c^3}.$$

So far we have not assumed $Z\alpha$ small. If we now make that assumption, and at the same time assume $D/(Z\alpha)$ small, we get on expanding (4.5.26) the approximation

$$-\frac{E}{hc} = Z^2 R \left[\frac{1}{(n + |j|)^2} + \frac{jD}{Z^3\alpha^3} + \frac{Z^2\alpha^2}{(n + |j|)^4} \left(\frac{n}{|j|} + \frac{1}{4} \right) \right], \qquad (4.5.27)$$

where R is the Rydberg constant

$$R = \frac{1}{2} \frac{mc^2\alpha^2}{hc} = \frac{2\pi^2 me^4}{h^3 c}. \qquad (4.5.28)$$

In this expression $\qquad \dfrac{D}{Z^3\alpha^3} = \dfrac{H\hbar^3}{m^2 c Z^2 e^3}. \qquad (4.5.29)$

In (4.5.27) the first term is the principal part, the second gives the Zeeman triplet, and the third gives the relativistic fine structure.[*]

Let us now consider the validity of the approximation used, assuming that $Z\alpha$ is small. Equation (4.4.24) tells us that ρ is of the order $(Z\alpha)^2$ and other orders are indicated in (4.4.23). Let us write out the square of (4.5.14), indicating below the orders of the terms:

$$[F(r)]^2 = -(B^2-1)\rho^2 + 2A\rho - (1-A^2) + DB \qquad -\frac{D^2}{4\rho^2} \qquad (4.5.30)$$
$$(Z\alpha)^2 \qquad (Z\alpha)^2 \quad (Z\alpha)^2 \quad D(Z\alpha)^{-1} \; D^2(Z\alpha)^{-4}.$$

It is clear then that our approximation based on the neglect of the last term is valid provided the quantity (4.5.29) is small compared with unity. The following special cases are of interest:

(i) If D is of the order $(Z\alpha)^6$ or higher power, then the Zeeman term in (4.5.27) is small compared with the relativistic fine structure.

(ii) If D is of the order $(Z\alpha)^5$, then the two terms are of the same order.

(iii) If D is of the order $(Z\alpha)^4$, then the Zeeman term predominates.

(iv) If D is of the order $(Z\alpha)^3$ or a lower power, then the method of approximation used above is not valid.

If we define $N = n + |j|$, we may write (4.5.27) in the form

$$-E = hcZ^2R\left[\frac{1}{N^2} + \frac{Z^2\alpha^2}{N^4}\left(\frac{N}{|j|} - \frac{3}{4}\right)\right] + j\frac{eH\hbar}{2mc}$$
$$(N = 1, 2, 3, \ldots; \; j = \pm 1, \pm 2, \pm 3, \ldots). \quad (4.5.31)$$

Thus the energy differs from that in the absence of the magnetic field by

$$j\frac{eH\hbar}{2mc} \quad (j = \pm 1, \pm 2, \pm 3, \ldots), \qquad (4.5.32)$$

and a transition from N to $N-1$ gives the Zeeman effect, a double line if j changes by -1 or $+1$ and a triple line if j changes by -1, 0 or $+1$ (cf. Sommerfeld, 1934, p. 321).

Remark. In Chapters II and III the object has been to develop a mathematical theory of rays and waves in space-time (relativistic

---

[*] Cf. Sommerfeld (1934, p. 259) for the relativistic fine structure term. The sign before this term is given incorrectly as minus by Bacher and Goudsmit (1932, pp. 218, 219).

geometrical mechanics) with emphasis on mathematical clarity rather than on physical meaning. In view of the optical analogy, it seems unlikely that we should get precisely correct physical results in all cases merely by imposing frequency and wave-length on top of this relativistic geometrical mechanics. We would expect correct results only when the wave-length was very small, in some appropriate relative sense. Nevertheless, as we have seen in §§ 4.4 and 4.5, the primitive quantization defined in § 4.3 does give the correct fine-structure energy formula and the simple Zeeman effect. Probably this is in the nature of a lucky chance. But we do get out of this the patterns of de Broglie waves in the hydrogenic atom, revealing symmetries which are not otherwise apparent. One might go on to discuss the Stark effect and also the Zeeman effect when the rays are not confined to a plane, as they are in the above treatment. These problems seem interesting mathematically, concerning as they do group properties of 3-waves associated with certain congruences of extremal world lines.

4.6. Interference for a single free particle and two holes

The most interesting ideal experiment illustrating the interference of material waves is that in which electrons pass through two holes or slits in a screen. We shall here discuss such interference patterns for two holes, treating first the case of a source-event and secondly the case of incident plane waves. Here, as in all the preceding theory, we are dealing with one moving particle, a system of rays representing possible histories of that particle. In Chapter v we shall deal with the problem of two particles.

Interference due to a source-event. Consider a screen fixed relative to a Galilean observer, dividing space into two regions. If we now open two holes in the screen, we have two regions of space-time, M' and M, representing the histories of these spatial regions, and between them the only connexion consists of two straight lines, H_1 and H_2, representing the histories of the holes (fig. 4.15).

A free particle of proper mass m can pass from M' to M through either of the holes. The appropriate medium-function is

$$f(x, \alpha) = mc(-\alpha_r \alpha_r)^{\frac{1}{2}}$$

as in (3.1.1), and there are *two* characteristic functions, one for refraction through each hole. By (3.2.11) they read

$$V_1(P', P) = mc[c^2(t - t')^2 - (r_1' + r_1)^2]^{\frac{1}{2}},$$
$$V_2(P', P) = mc[c^2(t - t')^2 - (r_2' + r_2)^2]^{\frac{1}{2}}. \qquad (4.6.1)$$

Here t' is the time of the initial event P' and r_1', r_2' are its distances from the holes; t, r_1, r_2 have the same meanings for the final event P.

We now take $P'(x')$ to be a given source-event and investigate the interference pattern behind the screen.

An increment h in V corresponds to an increment of 2π in phase angle, and so we get waves augmenting one another if $V_1 - V_2 = nh$ and destroying one another if $V_1 - V_2 = (n + \frac{1}{2}) h$, n being an integer. We are thus led to consider a continuous family of 3-spaces with equations

$$V_1 - V_2 = k, \qquad (4.6.2)$$

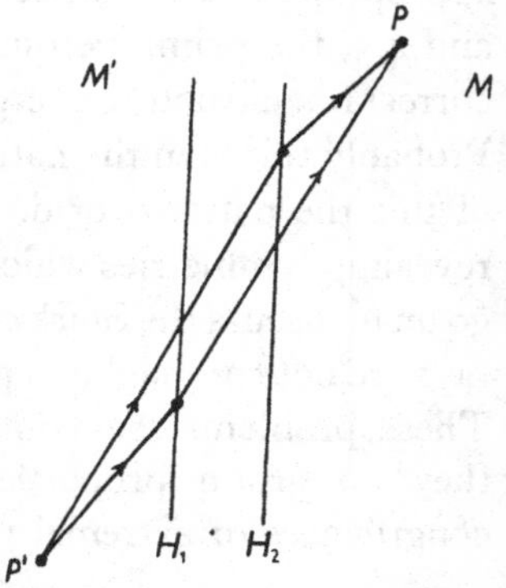

Fig. 4.15. Space-time diagram for refraction through two holes.

where k is a constant changing from 3-space to 3-space; these 3-spaces we shall call *interference 3-waves*. If we may use the convenient words *bright* and *dark* (of course without optical significance!) to indicate augmenting interference and destructive interference respectively, then we have

$$\text{for bright 3-waves,} \quad k = nh,$$
$$\text{for dark 3-waves,} \quad k = (n + \tfrac{1}{2}) h. \qquad (4.6.3)$$

Explicitly, the equations of the interference 3-waves are

$$mc[c^2(t - t')^2 - (r_1' + r_1)^2]^{\frac{1}{2}} - mc[c^2(t - t')^2 - (r_2' + r_2)^2]^{\frac{1}{2}} = k. \quad (4.6.4)$$

Here r_1, r_2 and t are running coordinates in space-time and r_1', r_2' and t' are fixed, since they belong to the source-event P'. This equation may be regarded as defining the histories of moving *interference 2-waves*, the instantaneous form of any one of these being obtained by putting $t = $ const. in (4.6.4). The continuous set of interference 2-waves include a certain discrete set of *bright 2-waves* and a certain discrete set of *dark 2-waves*, obtained by giving k the values in (4.6.3).

The 'zero' or 'central' bright 3-wave, corresponding to equal actions along the rays through the two holes, is given by putting $n = 0$; hence for it $k = 0$. Thus by (4.6.4) this central bright 3-wave is a fixed hyperboloidal surface making $r_1' + r_1 = r_2' + r_2$, or equivalently

$$r_2 - r_1 = r_1' - r_2'. \qquad (4.6.5)$$

This hyperboloid is bright for all values of t. It appears in fig. 4.16 as the hyperbola I having the holes for foci. For illustration we take $r_1' > r_2'$.

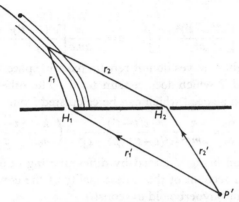

Fig. 4.16. Space picture of bright 2-waves produced from a source-event P' by two holes.

We consider h to be small. Then the bright 2-waves for $n = \pm 1$, ± 2, ... (but not too great) make $r_1' + r_1$ nearly equal to $r_2' + r_2$, since the right-hand side of (4.6.4) is small. Let us define ρ and ϵ by

$$\rho - \epsilon = r_1 + r_1', \quad \rho + \epsilon = r_2 + r_2', \qquad (4.6.6)$$

so that $\quad \rho = \tfrac{1}{2}(r_1 + r_2 + r_1' + r_2'), \quad \epsilon = \tfrac{1}{2}(r_2 - r_1 + r_2' - r_1'). \qquad (4.6.7)$

Then ρ and ϵ form a system of coordinates in space (an azimuthal angle is not required on account of symmetry). The equation of I is $\epsilon = 0$, and the bright 2-waves are to be found near I, i.e. for small values of ϵ. Accordingly we expand (4.6.4) in powers of ϵ and get

$$\frac{2\rho\epsilon}{[c^2(t-t')^2 - \rho^2]^{\frac{1}{2}}} = \frac{k}{mc}, \qquad (4.6.8)$$

or $$\epsilon = \frac{k}{2mc}\left[\frac{c^2(t-t')^2}{\rho^2} - 1\right]^{\frac{1}{2}}, \qquad (4.6.9)$$

or $$\rho = \frac{c(t-t')}{(1 + 4m^2c^2\epsilon^2/k^2)^{\frac{1}{2}}}. \qquad (4.6.10)$$

Of these forms, (4.6.9) is the most useful. To plot an interference 2-wave, we fix k and t. Now $\rho = $ const. is an ellipsoid of revolution having the holes for foci, and, if we assign ρ, then (4.6.9) tells us the value of ϵ, i.e. it tells us how far to go away from I (for which $\epsilon = 0$) on the ellipsoid $\rho = $ const., which of course cuts I orthogonally. Thus the bright 2-waves ($k = \pm h,\ \pm 2h,\ \ldots$) form a set of surfaces of revolution lying close to, and on both sides of, the hyperboloid I. For example, the bright 2-waves for $n = 1$ and $n = -1$ have the equations

$$\epsilon = \frac{h}{2mc}\left[\frac{c^2(t-t')^2}{\rho^2} - 1\right]^{\frac{1}{2}}, \quad \epsilon = -\frac{h}{2mc}\left[\frac{c^2(t-t')^2}{\rho^2} - 1\right]^{\frac{1}{2}}. \quad (4.6.11)$$

The bright 2-waves do not remain fixed in space, except for the hyperboloid I which does remain fixed. The others move away from it with a speed which can be calculated from

$$\frac{d\epsilon}{dt} = \frac{k}{2m\rho^2}\frac{c(t-t')}{[c^2(t-t')^2/\rho^2 - 1]^{\frac{1}{2}}} = \left(\frac{k}{2m\rho}\right)^2\frac{t-t'}{\epsilon}, \quad (4.6.12)$$

this equation being obtained by differentiating (4.6.9), holding ρ constant on account of the orthogonality of the confocal ellipsoid $\rho = $ const. and hyperboloid $\epsilon = $ const.

Interference for plane waves incident on two holes. The case of incident plane waves is easier than that of a source-event. Let us take the holes at positions with coordinates a_ρ and b_ρ. Then, as in (3.2.26), we have the two characteristic functions

$$\left.\begin{aligned}V_1(P) &= mc^2\gamma'\left(t - \frac{r_1 v'}{c^2} - \frac{v'_\rho a_\rho}{c^2} + \frac{C}{c^2}\right), \\ V_2(P) &= mc^2\gamma'\left(t - \frac{r_2 v'}{c^2} - \frac{v'_\rho b_\rho}{c^2} + \frac{C}{c^2}\right),\end{aligned}\right\} \quad (4.6.13)$$

where r_1 and r_2 are the distances of the final event P from the two holes, v'_ρ is the incident ray (or particle) velocity, and C is a constant depending on the choice of the particular incident 3-wave from which V is measured.

The interference 3-waves have the equations

$$V_1 - V_2 = m\gamma'[v'(r_2 - r_1) + v'_\rho(b_\rho - a_\rho)] = k, \quad (4.6.14)$$

and the bright and dark 3-waves are given by taking the values (4.6.3) for k.

We note that, whereas t was present in (4.6.4) and so gave us a moving pattern of interference 2-waves in the case of a source-event, t has disappeared from (4.6.14), and so we get for the interference 2-waves a fixed set of hyperboloids of revolution with the holes for foci:

$$r_2 - r_1 = -\frac{v'_\rho(b_\rho - a_\rho)}{v'} + \frac{k}{m\gamma'v'}. \qquad (4.6.15)$$

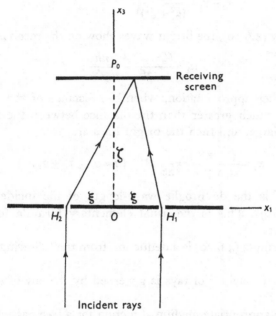

Fig. 4.17. Space picture for plane waves incident on two holes.

The bright 2-waves form the discrete set of confocal hyperboloids,

$$r_2 - r_1 = -\beta'_\rho(b_\rho - a_\rho) + \frac{nh}{m\gamma'v'} \quad (n = 0, \pm 1, \pm 2, \ldots), \quad (4.6.16)$$

where β'_ρ are the direction cosines of the incident rays.

To derive the simplest possible practical result, let us take the holes at $(\xi, 0, 0)$ and $(-\xi, 0, 0)$ and make the incident rays travel parallel to the positive sense of the x_3-axis (fig. 4.17). Then $\beta'_\rho(b_\rho - a_\rho) = 0$. Let a receiving screen be set up at $x_3 = \zeta$. Then on this screen we have

$$r_2 = [(x_1 + \xi)^2 + x_2^2 + \zeta^2]^{\frac{1}{2}}, \quad r_1 = [(x_1 - \xi)^2 + x_2^2 + \zeta^2]^{\frac{1}{2}}. \quad (4.6.17)$$

For points adjacent to $P_0(\zeta, o, o)$, x_1 and x_2 are small, and we have approximately

$$r_2 = (\xi^2 + \zeta^2)^{\frac{1}{2}} \left(1 + \frac{\xi x_1}{\xi^2 + \zeta^2} \right), \\
r_1 = (\xi^2 + \zeta^2)^{\frac{1}{2}} \left(1 - \frac{\xi x_1}{\xi^2 + \zeta^2} \right), \\
r_2 - r_1 = \frac{2\xi x_1}{(\xi^2 + \zeta^2)^{\frac{1}{2}}}, \quad (4.6.18)$$

and so, by (4.6.16), the bright waves show on the receiving screen as the lines

$$x_1 = \frac{(\xi^2 + \zeta^2)^{\frac{1}{2}}}{2\xi} \frac{nh}{m\gamma'v'}. \quad (4.6.19)$$

In a further approximation, when the distance of the receiving screen is much greater than the distance between the holes, we have ζ/ξ large, and then the bright lines are

$$x_1 = \frac{\zeta}{2\xi} \frac{nh}{m\gamma'v'} = \frac{\zeta}{2\xi} n\lambda' \quad (n = o, \pm 1, \pm 2, \ldots), \quad (4.6.20)$$

where λ' is the de Broglie wave-length of the incident waves (cf. (4.1.16)). This is the usual elementary formula for optical interference.

The formula (4.6.20) is a deduction from the following assumptions:

(i) The behaviour of rays is governed by the law of stationary action (2.1.3).

(ii) the appropriate medium-function for a free particle is as in (3.1.1).

(iii) The interference of 3-waves is governed by the law of primitive quantization contained in (4.3.1).

The formula gives us the loci on the receiving screen where waves from the two holes are in phase. We have called these lines *bright*, but that was only for dramatic effect, to make the loci easier to think about, since there is of course no question of optical brightness. Rather, in terms of current physical thought, we should interpret bright and dark loci in terms of *probability*. But since in primitive quantization we have made no attempt to follow the *amplitude* of a wave, we are in no position to make quantitative estimates of probability in general, although in the case of (4.6.20) it would be easy to make such an estimate, the amplitudes of the two interfering waves being regarded as constant in the small region of interference considered. But, taking the more cautious attitude and avoiding reference to amplitudes, we may conveniently change our language and say *very probable* where we said *bright*, and *very improbable* where we said *dark*. Thus if a particle arrives with velocity v' normal to the screen containing the holes, it is very probable that after passing through one or other of the two holes it will be found on one of the lines (4.6.20), and very improbable that it will be found on one of the lines midway between them.

CHAPTER V

SOME GENERALIZATIONS

5.1. Rays and waves in a Hamiltonian space

The theory so far developed in this book has had as its background the flat four-dimensional space-time of Minkowski with fundamental form $dx_r dx_r$. The most natural generalization would seem to be to the curved space-time of general relativity with fundamental form $g_{rs}(x) dx^r dx^s$. But it is wiser to go further in the direction of generalization and work in a manifold of N dimensions in which metric is not the basic concept. This is what we shall now do; the theory is, from a mathematical standpoint, a geometrization of the calculus of variations (cf. Hadamard, 1910; Carathéodory, 1935; Synge, 1951).

Consider a space of N dimensions with coordinates x^A; we shall use capital Latin suffixes for the range 1, 2, ..., N, with summation understood for repeated suffixes. The coordinates are real or purely imaginary, like Minkowskian time. But the reader who is troubled by the use of imaginary coordinates may think of the coordinates as real throughout this section, and later satisfy himself that it is permissible to use some purely imaginary coordinates, as is done in §5.2.

But whether we use real or imaginary coordinates, we must distinguish now between covariant and contravariant tensors, since we shall not restrict ourselves here to a limited group of transformations like the Lorentz transformations.

Let σ_A be the components of a covariant vector, and let an invariant relationship be assigned of the form

$$\Omega(\sigma, x) = 0. \tag{5.1.1}$$

This means that at any point x^A the components σ_A are not completely arbitrary; they must satisfy (5.1.1). We call σ_A the *slowness vector* and (5.1.1) the *slowness equation*.

The assignment of (5.1.1) makes our N-space a *Hamiltonian space* (H_N). The geometry of H_N consists in the properties of rays and waves in it, as defined below.

Take two points P', P in H_N and join them by a curve with equations

$$x^A = x^A(u), \tag{5.1.2}$$

u being a parameter. Let σ_A be assigned along the curve, so that we have

$$\sigma_A = \sigma_A(u), \tag{5.1.3}$$

the assignment being made to satisfy (5.1.1) at each point of the curve. Then the integral

$$I = -\int_{P'}^{P} \sigma_A \, \mathrm{d}x^A \tag{5.1.4}$$

has some value, depending on the choice of the points P', P, of the curve joining them, and of the way in which σ_A has been assigned.

We define a *ray* in H_N as a curve which gives a stationary value to I, the end points being held fixed and (5.1.1) being satisfied for all the curves considered. Thus a ray is defined by

$$\delta \int \sigma_A \, \mathrm{d}x^A = 0, \quad \Omega(\sigma, x) = 0. \tag{5.1.5}$$

By the same reasoning as that following (2.1.16), we obtain the differential equations of a ray in Hamiltonian form:

$$\frac{\mathrm{d}x^A}{\mathrm{d}w} = \frac{\partial \Omega}{\partial \sigma_A}, \quad \frac{\mathrm{d}\sigma_A}{\mathrm{d}w} = -\frac{\partial \Omega}{\partial x^A}, \tag{5.1.6}$$

w being some particular parameter along the ray. It is related to the increments in the coordinates by

$$\sigma_A \, \mathrm{d}x^A = \sigma_A \frac{\partial \Omega}{\partial \sigma_A} \, \mathrm{d}w. \tag{5.1.7}$$

We have thus set up in H_N the elements of a geometry, if we regard the rays as 'straight lines' and the integral I, taken along a ray, as a measure of 'distance'. However, the word 'distance' may be needed in other physical senses and it is better to avoid it in this connexion. We shall use the word *action* instead, defining the action from P' to P along the ray joining them to be

$$V(P', P) = -\int_{P'}^{P} \sigma_A \, \mathrm{d}x^A. \tag{5.1.8}$$

We also call V the *characteristic function* of H_N. The insertion of the minus sign in (5.1.4) and (5.1.8) is convenient in relativistic theory.

We note that a ray is determined by assigning either (a) two points on it (by the stationary principle), or (b) initial values of x^4 and σ_A, subject to $\Omega(\sigma, x) = 0$ (by the differential equations (5.1.6)).

Determination of the action element or medium-function. In accordance with (5.1.8) the element of action on a ray is

$$dI = - \sigma_A dx^A, \tag{5.1.9}$$

the vector σ_A being determined along the ray by the equations (5.1.6). Suppose, however, that we wish to define an action element on a curve C which is not a ray, the definition being such that it reduces to (5.1.9) when C is a ray. As it stands, (5.1.9) does not suffice as a definition, because we do not know what values to insert for σ_A. Even if we impose the condition $\Omega(\sigma, x) = 0$, we are left with a wide liberty of choice in σ_A.

To make the element of action (5.1.9) meaningful for any curve C, we impose on σ_A the conditions

$$dx^A = \frac{\partial \Omega}{\partial \sigma_A} dw, \quad \Omega(\sigma, x) = 0. \tag{5.1.10}$$

(Note that the first of these is one half of the Hamiltonian equations (5.1.6).) Now, given dx^A, an infinitesimal displacement on C, we have in (5.1.10) $N+1$ equations which suffice to determine the $N+1$ quantities σ_A, dw. In this way we associate a definite vector σ_A with each point of any curve C, at the same time determining a parametric element dw. Thus the action element is determined by x^A and dx^A, and we may write for it

$$f(x, dx) = - \sigma_A dx^A = - \sigma_A \frac{\partial \Omega}{\partial \sigma_A} dw. \tag{5.1.11}$$

Now if instead of taking the element dx^A on the curve C, we take $\overline{d}x^A = k\,dx^A$ where k is a positive factor, then the equations (5.1.10) give us $\overline{d}w = k\,dw$ and the same values of σ_A as before. Hence

$$f(x, \overline{d}x) = - \sigma_A \overline{d}x^A,$$

$$f(x, k\,dx) = kf(x, dx) \quad (k > 0),$$

and so the function $f(x, dx)$ is positive homogeneous of degree unity in the differentials.

If, along any curve C, we attach vectors σ_A in accordance with (5.1.10), we shall call C *a curve with attached slowness.*

To sum up (and this process is fundamental) *the action element* $f(x, dx)$ *is determined from the function* $\Omega(\sigma, x)$ *by eliminating the* $N+1$ *quantities* σ_A, dw *from the* $N+2$ *equations*

$$dx^A = \frac{\partial \Omega}{\partial \sigma_A} dw, \quad f = -\sigma_B \frac{\partial \Omega}{\partial \sigma_B} dw, \quad \Omega(\sigma, x) = 0, \quad (5.1.12)$$

and solving for f the resulting equation connecting f, x^A, dx^A.

There is no necessity to use infinitesimal notation. The same *function $f(x, X)$* (we shall call it the *medium-function*) is obtained by eliminating σ_A, θ from the equations

$$X^A = \theta \frac{\partial \Omega}{\partial \sigma_A}, \quad f = -\theta \sigma_B \frac{\partial \Omega}{\partial \sigma_B}, \quad \Omega(\sigma, x) = 0, \quad (5.1.13)$$

and solving for f. The resulting function is positive homogeneous of degree unity in X^A:

$$f(x, kX) = kf(x, X) \quad (k > 0). \quad (5.1.14)$$

Rays in terms of $f(x, dx)$. Consider a ray joining P' to P. Consider all curves with attached slowness adjacent to the ray and passing through the same end points. Then along each of these curves there is an integral

$$I = -\int_{P'}^{P} \sigma_A dx^A = \int_{P'}^{P} f(x, dx). \quad (5.1.15)$$

Now, by definition, the ray satisfies $\delta I = 0$ when on the varied curves σ_A is arbitrary except for $\Omega(\sigma, x) = 0$. Hence $\delta I = 0$ for the more restricted choice of σ_A, viz. that satisfying (5.1.10). *Therefore a ray satisfies the variational principle*

$$\delta \int_{P'}^{P} f(x, dx) = 0, \quad (5.1.16)$$

or

$$\delta \int_{P'}^{P} f(x, X) dw = 0, \quad (5.1.17)$$

where $X^A = dx^A/dw$. Hence the rays satisfy the Euler-Lagrange equations

$$\frac{d}{dw} \frac{\partial f(x, X)}{\partial X^A} - \frac{\partial f(x, X)}{\partial x^A} = 0. \quad (5.1.18)$$

In these equations w is any monotonically increasing parameter, and not necessarily the special parameter of the Hamiltonian equations (5.1.6).

The slowness equation $\Omega(\sigma, x) = 0$ derived from the function $f(x, X)$.
Consider any curve C with attached slowness σ_A, obtained from
(5.1.10). Now C may be drawn through any point in any direction,
and moreover it may be parametrized arbitrarily. Although the
special parameter w was present in (5.1.10), we now deliberately
choose an arbitrary parameter u. We can then declare that the $2N$
quantities

$$x^A, \quad X^A = \frac{dx^A}{du}, \qquad (5.1.19)$$

are arbitrary and independent.

On account of the homogeneity of f, we have from (5.1.11)

$$f(x, X) = -\sigma_A X^A, \qquad (5.1.20)$$

and we have also $\qquad \Omega(\sigma, x) = 0. \qquad (5.1.21)$

Further, by (5.1.10), allowing for the change of parameter by the
undetermined factor θ, we have

$$X^A = \theta \frac{\partial \Omega}{\partial \sigma_A}. \qquad (5.1.22)$$

We recognize in (5.1.20), (5.1.21) and (5.1.22) $N + 2$ equations
involving the quantities

$$\sigma_A, \quad \theta, \quad x^A, \quad X^A,$$

and of these the $2N$ quantities (5.1.19) are arbitrary and independent.

Varying these independent quantities, and consequently σ_A,
we get from (5.1.20), with use of (5.1.22),

$$\frac{\partial f}{\partial X^A} \delta X^A + \frac{\partial f}{\partial x^A} \delta x^A = -\sigma_A \delta X^A - X^A \delta \sigma_A$$

$$= -\sigma_A \delta X^A - \theta \frac{\partial \Omega}{\partial \sigma_A} \delta \sigma_A. \qquad (5.1.23)$$

But by variation of (5.1.21) we have

$$\frac{\partial \Omega}{\partial \sigma_A} \delta \sigma_A + \frac{\partial \Omega}{\partial x^A} \delta x^A = 0, \qquad (5.1.24)$$

and so (5.1.23) may be written

$$\left(\frac{\partial f}{\partial X^A} + \sigma_A \right) \delta X^A + \left(\frac{\partial f}{\partial x^A} - \theta \frac{\partial \Omega}{\partial x^A} \right) \delta x^A = 0. \qquad (5.1.25)$$

Here, by (5.1.19), the variations are arbitrary and independent, and so we have the important relations between the slowness vector and the derivatives of the medium-function

$$\sigma_A = -\frac{\partial f(x, X)}{\partial X^A}, \qquad (5.1.26)$$

and the less important relations

$$\frac{\partial f(x, X)}{\partial x^A} = \theta \frac{\partial \Omega(\sigma, x)}{\partial x^A}. \qquad (5.1.27)$$

Since f is homogeneous of degree unity in the X's, the right-hand side of (5.1.26) is homogeneous of degree zero; hence it is independent of the parametrization u employed, since the effect of changing the parametrization is simply to multiply the X's by a common factor.

The equation (5.1.26) enables us to calculate the slowness vector at once when we are given the medium-function. But it does more. Since the right-hand sides are homogeneous of degree zero in the X's, they are functions only of their $N-1$ ratios; thus these $N-1$ quantities can be eliminated from the N equations, and the result is a relation between σ_A and x^A, which is precisely the equation $\Omega(\sigma, x) = 0$. *Thus by eliminating X^A from (5.1.26) we obtain the slowness equation $\Omega(\sigma, x) = 0$, assuming that the medium-function $f(x, X)$ is given.*

Invariance under coordinate transformations. We think of general transformations of the coordinates x^A in H_N. We have already stipulated that σ_A is a covariant vector and $\Omega(\sigma, x)$ an invariant. Then at a fixed point of H_N, the variation $\delta\sigma_A$ is a covariant vector and the variation $\delta\Omega$ is an invariant. But (since $\delta x^A = 0$)

$$\delta\Omega = \frac{\partial \Omega}{\partial \sigma_A} \delta\sigma_A, \qquad (5.1.28)$$

and hence, by the well-known test for tensor character, we see that $\partial\Omega/\partial\sigma_A$ is a contravariant vector. Then by the first of (5.1.12) dx^A/dw is a contravariant vector; but dx^A is itself a contravariant vector, and so dw is an invariant. Then the second of (5.1.12) tells us that the action element $f(x, dx)$ is an invariant. The medium-function $f(x, X)$, where $X^A = dx^A/du$, is then obviously an invariant also.

Thus, starting from an equation $\Omega(\sigma, x) = 0$ in which Ω is invariant, we can obtain an invariant differential form $f(x, dx)$, positive homogeneous of degree unity in the differentials. A geometry based on such a differential form is called a *Finsler geometry*, Finsler geometries including Riemannian geometry as a special case and Euclidean geometry as a still more special case. What we have done above may be described by saying that *we have established the equivalence of Hamiltonian geometry and Finsler geometry;* a Hamiltonian space is a Finsler space, and conversely, and we have given a procedure for passing from one way of looking at the space to the other ($\Omega = 0 \to f$ and $f \to \Omega = 0$).

Summary. To sum up, a Hamiltonian space is characterized by a slowness equation

$$\Omega(\sigma, x) = 0, \tag{5.1.29}$$

where Ω is an invariant function of a covariant slowness vector σ_A and the coordinates x^A. Rays satisfy

$$\delta \int \sigma_A \, dx^A = 0, \quad \Omega(\sigma, x) = 0, \tag{5.1.30}$$

and have the equations

$$\frac{dx^A}{dw} = \frac{\partial \Omega}{\partial \sigma_A}, \quad \frac{d\sigma_A}{dw} = -\frac{\partial \Omega}{\partial x^A}, \tag{5.1.31}$$

for a special parameter w. An invariant action element (or equivalently medium-function) $f(x, dx)$ is defined by

$$dx^A = \frac{\partial \Omega}{\partial \sigma_A} \, dw, \quad f(x, dx) = -\sigma_B \frac{\partial \Omega}{\partial \sigma_B} \, dw, \quad \Omega(\sigma, x) = 0, \tag{5.1.32}$$

and the rays satisfy $\qquad \delta \int f(x, dx) = 0. \tag{5.1.33}$

The element of action may be written

$$f(x, dx) = -\sigma_A \, dx^A, \tag{5.1.34}$$

and the characteristic function, or action along a ray, is

$$V(P', P) = -\int_{P'}^{P} \sigma_A \, dx^A = \int_{P'}^{P} f(x, dx). \tag{5.1.35}$$

For variation of the end points of a ray,

$$\delta V = -\sigma_A \delta x^A + \sigma'_A \delta x'^A, \tag{5.1.36}$$

so that $\qquad \dfrac{\partial V}{\partial x^A} = -\sigma_A, \quad \dfrac{\partial V}{\partial x'^A} = \sigma'_A, \tag{5.1.37}$

and V satisfies the Hamilton-Jacobi equation

$$\Omega\left(-\frac{\partial V}{\partial x}, x\right) = 0. \tag{5.1.38}$$

Waves in H_N. To define waves in H_N, we start from an initial wave, say Σ_0, a subspace of $N-1$ dimensions with equation $F(x) = 0$. Now, in an N-space without metric, an $(N-1)$-space has no contravariant normal, but it has a covariant normal; for any displacement dx^A in $F(x) = 0$ we have

$$F_A \, dx^A = 0, \tag{5.1.39}$$

where $F_A = \partial F / \partial x^A$, and so we call the covariant vector F_A the normal to $F(x) = 0$. Choose then a slowness vector σ'_A on Σ_0 according to $\sigma'_A = \theta F_A$ and find θ so that $\Omega(\sigma', x') = 0$ is satisfied; thus we have for θ the equation

$$\Omega(\theta F, x') = 0, \tag{5.1.40}$$

where in this brief notation F stands for its partial derivatives and x' is a point on Σ_0. Draw through the points of Σ_0 rays, as given by (5.1.31), using the values of σ' and x' just considered as initial values, the rays starting with directions in H_N given by

$$\frac{dx'_r}{dw} = \frac{\partial \Omega(\sigma', x')}{\partial \sigma'_A}. \tag{5.1.41}$$

Then measure off on these rays an action C, the same for all of them, thus generating an $(N-1)$-*wave* Σ with the equation

$$V(P', P) = -\int_{P'}^{P} \sigma_A \, dx^A = C, \tag{5.1.42}$$

where P' is on Σ_0 and P on Σ. By varying C we get a single infinity of waves. It is easy to see that these waves have the group property: the same set of waves is constructed by starting from any one of them.

Having so defined $(N-1)$-waves in H_N (they are *transversals* in the language of the calculus of variations), we can carry over into H_N the primitive quantization of Chapter IV. This means that two adjacent $(N-1)$-waves with the same phase have between them an action separation

$$V(P', P) = -\int_{P'}^{P} \sigma_A \, dx^A = h, \tag{5.1.43}$$

where h is Planck's constant, the integral being taken along a ray from wave to wave.

Enough has now been said about a Hamiltonian N-space to enable the reader to translate into these more general terms ideas developed in previous chapters against the background of Minkowskian space-time, but not actually dependent on that background. We shall now proceed with specific applications.

5.2. The two-body problem

Consider two Euclidean planes, Π and $\overline{\Pi}$, with rectangular Cartesian coordinates (x, y) in Π and $(\bar{x}, \bar{y})$ in $\overline{\Pi}$. An ordered pair of points, P in Π and $\bar{P}$ in $\overline{\Pi}$, provides us with four numbers $(x, y, \bar{x}, \bar{y})$. These four numbers may be regarded as defining a 'point' in a 4-space, which is in fact the 'product' of the planes Π and $\overline{\Pi}$ in a topological sense.

Let us carry over this plan into relativity. Consider two particles in the flat space-time of Minkowski; let x_r be an event in the history of one particle and $\bar{x}_r$ an event in the history of the other. We shall use the same space-time axes for both the particles. Then the product of space-time by itself is an 8-space, a point of the 8-space having the eight coordinates $(x_r, \bar{x}_r)$ or briefly $(x, \bar{x})$. We recall that small Latin suffixes have the range 1, 2, 3, 4 and $x_4 = ict$, $\bar{x}_4 = ic\bar{t}$.

Our purpose is to develop a Hamiltonian geometry in this 8-space, and this is done, as in § 5.1, by writing down an invariant slowness equation
$$\Omega(\sigma, \overline{\sigma}, x, \bar{x}) = 0, \tag{5.2.1}$$
where σ, $\overline{\sigma}$ stand for a set of eight quantities σ_r, $\overline{\sigma}_r$ (slowness components). For relativistic invariance, we demand that Ω shall be invariant in form under the application of the same Lorentz transformation to x_r and $\bar{x}_r$, σ_r and $\overline{\sigma}_r$ transforming as two 4-vectors.

In order to visualize the discussion, we may think in three different ways:

(i) We may visualize in H_8, a ray appearing as a curve and a wave as a space of seven dimensions.

(ii) We may visualize in space-time, a point of H_8 appearing as a pair of events (one for each particle) and a ray appearing as a pair of world lines, with their events put in correspondence by the parameter w of (5.1.31).

(iii) We may form kinematical pictures in the space of a Galilean observer, but this is apt to be confusing because corresponding events will in general have different times.

Let us now make the following special choice of slowness equation for the case of two particles subject to no external field:

$$\Omega = \frac{1}{2m}\,\sigma_r\,\sigma_r + \frac{1}{2\overline{m}}\,\overline{\sigma}_r\overline{\sigma}_r + F(x,\overline{x}) + \tfrac{1}{2}(m+\overline{m})\,c^2 = 0. \quad (5.2.2)$$

Here m and $\overline{m}$ are the proper masses of the two particles and $F(x,\overline{x})$ some function of their eight coordinates, representing an interaction potential.

From a mathematical standpoint we are at liberty to take any slowness equation and see what comes out of it, for the physical justification of any such formula is to be found in the success of physical predictions based on it, if they are successful. But if we decide to investigate two-body problems along these lines, it is hard to see what equation to adopt other than (5.2.2). For (3.1.6) seems to be the proper slowness equation for a single free particle, and (5.2.2) is the immediate and natural generalization for the case of two particles. If we drop the interaction term from (5.2.2) by putting $F=0$, and then regard σ_r and $\overline{\sigma}_r$ as the momentum-energy 4-vectors of the two particles, we have by (3.1.6)

$$\sigma_r\,\sigma_r = -m^2c^2, \quad \overline{\sigma}_r\overline{\sigma}_r = -\overline{m}^2c^2,$$

and so (5.2.2) is satisfied. This does not of course *prove* that (5.2.2) is the proper formula to take, but it does give it sufficient standing to make it worth investigating.

To find the element of action corresponding to (5.2.2), we use (5.1.32) with appropriate change of notation. We have

$$\begin{aligned}
\mathrm{d}x_r &= \frac{\sigma_r}{m}\,\mathrm{d}w, \quad \mathrm{d}\overline{x}_r = \frac{\overline{\sigma}_r}{\overline{m}}\,\mathrm{d}w, \\
f &= -\left(\frac{\sigma_r\,\sigma_r}{m} + \frac{\overline{\sigma}_r\overline{\sigma}_r}{\overline{m}}\right)\mathrm{d}w \\
&= [2F + (m+\overline{m})\,c^2]\,\mathrm{d}w.
\end{aligned} \qquad (5.2.3)$$

Hence
$$\begin{aligned}
\sigma_r &= m\frac{\mathrm{d}x_r}{\mathrm{d}w}, \quad \overline{\sigma}_r = \overline{m}\frac{\mathrm{d}\overline{x}_r}{\mathrm{d}w}, \\
\frac{1}{m}\,\sigma_r\,\sigma_r &= \frac{m\,\mathrm{d}x_r\,\mathrm{d}x_r}{\mathrm{d}w^2}, \quad \frac{1}{\overline{m}}\,\overline{\sigma}_r\overline{\sigma}_r = \frac{\overline{m}\,\mathrm{d}\overline{x}_r\,\mathrm{d}\overline{x}_r}{\mathrm{d}w^2},
\end{aligned} \qquad (5.2.4)$$

and so (5.2.2) gives for dw the equation

$$m\,dx_r\,dx_r + \overline{m}\,d\overline{x}_r\,d\overline{x}_r + [2F + (m + \overline{m})\,c^2]\,dw^2 = 0. \qquad (5.2.5)$$

Thus $\quad dw = [2F + (m + \overline{m})\,c^2]^{-\frac{1}{2}}[-m\,dx_r\,dx_r - \overline{m}\,d\overline{x}_r\,d\overline{x}_r]^{\frac{1}{2}}, \qquad (5.2.6)$

and the required action element is, by (5.2.3),

$$f(x, \overline{x}, dx, d\overline{x}) = [2F + (m + \overline{m})\,c^2]^{\frac{1}{2}}[-m\,dx_r\,dx_r - \overline{m}\,d\overline{x}_r\,d\overline{x}_r]^{\frac{1}{2}}. \qquad (5.2.7)$$

This represents the transformation of the Hamiltonian space defined by (5.2.2) into a Finsler space.

The equations of the rays are, by (5.1.31),

$$\left. \begin{array}{ll} \dfrac{dx_r}{dw} = \dfrac{\sigma_r}{m}, & \dfrac{d\sigma_r}{dw} = -\dfrac{\partial F}{\partial x_r}, \\[2mm] \dfrac{d\overline{x}_r}{dw} = \dfrac{\overline{\sigma}_r}{\overline{m}}, & \dfrac{d\overline{\sigma}_r}{dw} = -\dfrac{\partial F}{\partial \overline{x}_r}. \end{array} \right\} \qquad (5.2.8)$$

We note that, from its definition in terms of a variational principle, a ray is determined when we are given a pair of initial events P', $\overline{P}'$ (one for each particle) and a pair of final events P, $\overline{P}$ (one for each particle). Also it is clear from the form of (5.2.8) that a ray is determined by initial values of x_r, $\overline{x}_r$, σ_r, $\overline{\sigma}_r$, i.e. by a pair of initial events with slowness 4-vectors associated with them, these sixteen quantities being arbitrary except for the single condition $\Omega = 0$. These initial data determine the initial velocities of the two particles, for we have

$$\frac{v_\rho}{ic} = \frac{dx_\rho}{dx_4} = \frac{\sigma_\rho}{\sigma_4}, \qquad \frac{\overline{v}_\rho}{ic} = \frac{d\overline{x}_\rho}{d\overline{x}_4} = \frac{\overline{\sigma}_\rho}{\overline{\sigma}_4}, \qquad (5.2.9)$$

where v_ρ and $\overline{v}_\rho$ are the velocities in question.

But a pair of initial events with initial velocities do not suffice to determine a ray, for they give only the ratios σ_ρ/σ_4 and $\overline{\sigma}_\rho/\overline{\sigma}_4$, and there is still one degree of freedom after making use of $\Omega = 0$. In fact, if we are given the initial velocities, we may regard σ_ρ and $\overline{\sigma}_\rho$ as known in terms of σ_4 and $\overline{\sigma}_4$, these last being free except for the equation

$$\frac{1}{2m}\left(1 - \frac{v^2}{c^2}\right)\sigma_4^2 + \frac{1}{2\overline{m}}\left(1 - \frac{\overline{v}^2}{c^2}\right)\overline{\sigma}_4^2 + F(x, \overline{x}) + \tfrac{1}{2}(m + \overline{m})\,c^2 = 0.$$
$$\qquad (5.2.10)$$

It is interesting to compare and contrast the general conclusions of Newtonian mechanics and of Hamiltonian geometrical mechanics as developed here. In both, the history of a pair of particles is determined if we know a pair of initial events and a pair of final events. But whereas initial events and velocities determine the motion of a pair of particles in Newtonian mechanics, in the present theory they do not; there is one undetermined parameter.

The hydrogenic atom. In a recent paper (Synge, 1953) the above method has been applied to the hydrogenic atom, using (5.2.2) with the interaction potential

$$F(x, \bar{x}) = e\bar{e}[(x_r - \bar{x}_r)(x_r - \bar{x}_r)]^{-\frac{1}{2}}, \qquad (5.2.11)$$

$e, \bar{e}$ being the electric charges on the particles and the space-time displacement $x_r - \bar{x}_r$ between corresponding events (i.e. events having the same parametric value w) being assumed space-like. The proper mass M of the system of two particles is suitably defined, and the quantized levels for its square are found to be (accurately in terms of the method of primitive quantization)

$$M^2 = (m + \bar{m})^2 - \frac{e^2\bar{e}^2}{c^2\hbar^2}\frac{m\bar{m}}{(n+j)^2} \quad (n = 0, 1, 2, \ldots; j = 1, 2, \ldots). \quad (5.2.12)$$

If $Z\alpha$ is small, this gives the following approximate levels for the proper energy of the system:

$$Mc^2 = (m + \bar{m})c^2 - \frac{1}{2}\frac{m\bar{m}}{m+\bar{m}}\frac{Z^2\alpha^2c^2}{(n+j)^2}, \qquad (5.2.13)$$

where
$$\bar{e} = -Ze, \quad \alpha = \frac{e^2}{c\hbar}. \qquad (5.2.14)$$

Thus the mass correction factor comes out correctly, no approximation being made with regard to the ratio $m/\bar{m}$ of the masses.

For further details the reader is referred to the paper cited.

5.3. Two free particles in empty space-time

To deal with two free particles, we put $F = 0$ in (5.2.2) and obtain

$$\Omega = \frac{1}{2m}\sigma_r\sigma_r + \frac{1}{2\bar{m}}\bar{\sigma}_r\bar{\sigma}_r + \frac{1}{2}(m+\bar{m})c^2 = 0. \qquad (5.3.1)$$

By (5.2.7) the action element is then

$$f(dx, d\bar{x}) = c(m+\bar{m})^{\frac{1}{2}}(-m\,dx_r\,dx_r - \bar{m}\,d\bar{x}_r\,d\bar{x}_r)^{\frac{1}{2}}. \qquad (5.3.2)$$

By (5.2.8) we get for a ray

$$\frac{dx_r}{dw} = \frac{\sigma_r}{m} = \text{const.}, \quad \frac{d\bar{x}_r}{dw} = \frac{\bar{\sigma}_r}{\bar{m}} = \text{const.} \qquad (5.3.3)$$

Thus a ray gives us two straight lines in space-time, the velocities on them being constants:

$$v_\rho = \mathrm{i}c\,\frac{\sigma_\rho}{\sigma_4}, \quad \bar{v}_\rho = \mathrm{i}c\,\frac{\overline{\sigma}_\rho}{\overline{\sigma}_4}. \tag{5.3.4}$$

We have the equation (5.2.10) with $F = 0$.

We may refer to $(x_r, \bar{x}_r)$ or briefly $(x, \bar{x})$ as an *event-pair* $(P, \bar{P})$. Now if we take an initial event-pair $(P', \bar{P}')$ and a final event-pair $(P, \bar{P})$, they determine a ray in H_8, and the action along it is

$$V(P', \bar{P}'; P, \bar{P}) = \int f(\mathrm{d}x, \mathrm{d}\bar{x}) = wc(m + \overline{m})^{\frac{1}{2}}\left(-\frac{\sigma_r\sigma_r}{m} - \frac{\overline{\sigma}_r\overline{\sigma}_r}{m}\right)^{\frac{1}{2}},$$
$$\tag{5.3.5}$$

where $w = \int \mathrm{d}w$. But by (5.3.3)

$$\left.\begin{aligned} x_r - x'_r &= w\,\frac{\sigma_r}{m}, \quad \bar{x}_r - \bar{x}'_r = w\,\frac{\overline{\sigma}_r}{m}, \\ w^2\left(\frac{\sigma_r\sigma_r}{m} + \frac{\overline{\sigma}_r\overline{\sigma}_r}{\overline{m}}\right) &= m(x_r - x'_r)(x_r - x'_r) + \overline{m}(\bar{x}_r - \bar{x}'_r)(\bar{x}_r - \bar{x}'_r), \end{aligned}\right\} \tag{5.3.6}$$

and so

$$V = c(m + \overline{m})^{\frac{1}{2}}\left[-m(x_r - x'_r)(x_r - x'_r) - \overline{m}(\bar{x}_r - \bar{x}'_r)(\bar{x}_r - \bar{x}'_r)\right]^{\frac{1}{2}}, \tag{5.3.7}$$

as indeed is also easily seen from (5.3.2). *This is the characteristic function for a pair of free particles in empty space-time.*

If we assign only an initial event-pair $(P', \bar{P}')$ and consider the totality of rays in H_8 coming from this point, we have the analogue of the source-event for a single particle. We may call it a *source-event-pair*. A source-event-pair fills H_8 with 7-waves, with the equations

$$V(x', \bar{x}'; x, \bar{x}) = C, \tag{5.3.8}$$

wherein x'_r, $\bar{x}'_r$ are constants (pertaining to the source-event-pair) and x_r, $\bar{x}_r$ are current coordinates.

By (5.3.7) the 7-waves may also be exhibited in the form

$$m(x_r - x'_r)(x_r - x'_r) = -\overline{m}(\bar{x}_r - \bar{x}'_r)(\bar{x}_r - \bar{x}'_r) - \frac{V^2}{(m + \overline{m})c^2}, \tag{5.3.9}$$

V being a constant for each of them. This enables us to get some sort of a picture in space-time of the 7-waves. If we assign values to $\bar{x}$ and V, and treat x as running coordinates, (5.3.9) gives a pseudosphere in space-time with centre at x'. If we change $\bar{x}$ and/or V, we get a new pseudosphere with the same centre and in general a different radius. But the representation is imperfect since we get the same pseudosphere by changing $\bar{x}$ and V in such a way as to leave the right-hand side of (5.3.9) unchanged.

Plane 7-waves. To generate a system of plane 7-waves, we start with a 7-dimensional plane Σ_0 in H_8 with equation

$$A_r x_r + \bar{A}_r \bar{x}_r = C, \qquad (5.3.10)$$

A_r, $\bar{A}_r$ and C being any constants. The appropriate values of the initial components of slowness are

$$\sigma_r = \theta A_r, \quad \bar{\sigma}_r = \theta \bar{A}_r \qquad (5.3.11)$$

(cf. (5.1.39)), where θ is to be found by substituting in $\Omega = 0$ (5.3.1). This gives $\theta = $ const. (the value is of no importance), and so, since σ_r and $\bar{\sigma}_r$ are constants along the rays, we may write the equations of any set of plane 7-waves in the form

$$\sigma_r x_r + \bar{\sigma}_r \bar{x}_r = \text{const.}, \qquad (5.3.12)$$

the corresponding rays being given by

$$\frac{dx^r}{dw} = \frac{\partial\Omega}{\partial\sigma_r} = \frac{\sigma_r}{m}, \quad \frac{d\bar{x}_r}{dw} = \frac{\partial\Omega}{\partial\bar{\sigma}_r} = \frac{\bar{\sigma}_r}{\bar{m}}. \qquad (5.3.13)$$

Here σ_r and $\bar{\sigma}_r$ are constants of the system. The characteristic function of the system is

$$V(P, \bar{P}) = -\sigma_r x_r - \bar{\sigma}_r \bar{x}_r + \text{const.}, \qquad (5.3.14)$$

the value of the constant depending on which of the 7-waves we choose to be $V = 0$.

The totality of rays corresponding to these plane 7-waves form an ∞^7 set of parallel straight lines in H_8. In the space-time picture each such ray gives us a pair of straight world lines traversed with constant velocities

$$v_\rho = ic\frac{\sigma_\rho}{\sigma_4}, \quad \bar{v}_\rho = ic\frac{\bar{\sigma}_\rho}{\bar{\sigma}_4}. \qquad (5.3.15)$$

To appreciate an indeterminacy present in the case of two particles, let us return for a moment to the case of a single free particle, for which we have, as in (3.1.5),

$$\sigma_\rho = m\gamma v_\rho, \quad \sigma_4 = im\gamma c, \quad \gamma = \left(1 - \frac{v^2}{c^2}\right)^{-\frac{1}{2}}.$$

If we are told that the single particle has a certain constant velocity v_ρ, but have no knowledge of its position, we think naturally of a system of plane 3-waves in space-time with equations

$$\sigma_r x_r = \text{const.,}$$

the values of σ_r being known in terms of v_ρ as above. But if we have *two* free particles, and are told that their velocities are constant with values v_ρ, $\overline{v}_\rho$, we cannot obtain σ_r, $\overline{\sigma}_r$ from (5.3.15) and $\Omega = 0$, which, as in (5.2.10), reads

$$\frac{1}{2m}\left(1 - \frac{v^2}{c^2}\right)\sigma_4^2 + \frac{1}{2\overline{m}}\left(1 - \frac{\overline{v}^2}{c^2}\right)\overline{\sigma}_4^2 + \tfrac{1}{2}(m + \overline{m})c^2 = 0. \quad (5.3.16)$$

We are one equation short, and so there is still a degree of freedom in the choice of σ_r, σ_r, as has been remarked earlier in a different connexion. This means that the picture of two free particles moving with constant velocities v_ρ, $\overline{v}_\rho$, but with positions unspecified, cannot summon up in our minds the picture of a definite system of plane 7-waves in H_8. Something more must be added.

Now, as we have seen already, an initial event-pair $(P', \overline{P}')$ together with a final event-pair $(P, \overline{P})$ provide us with a definite characteristic function V. This suggests that we may get definite plane 7-waves by allowing P' and $\overline{P}'$ to recede to infinity in a definite way, just as in optics we get plane waves by removing the light-source to

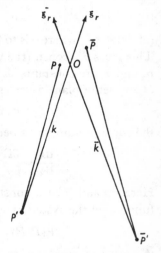

Fig. 5.1. Space-time diagram of the recession of initial events P', $\overline{P}'$ into the remote past. $P'P$ and $\overline{P}'\overline{P}$ together represent a ray in H_8.

infinity. Let us then (fig. 5.1) hold P, $\overline{P}$ fixed and let P', $\overline{P}'$ recede into the remote past by moving them in space-time in the following manner. Taking two time-like unit 4-vectors, $\overline{\xi}_r$, ξ_r, both pointing into the future, we choose as initial events

$$P': \quad x_r' = -k\xi_r, \qquad \overline{P}': \quad \overline{x}_r' = -\overline{k}\overline{\xi}_r, \qquad (5.3.17)$$

where k, $\overline{k}$ are any positive numbers, being in fact the Minkowskian distances $P'O$, $\overline{P}'O$, where O is the origin of space-time. We let k and $\overline{k}$ tend to infinity in such a way that the ratio $\overline{k}/k$ tends to a finite limit ρ. Then, by (5.3.7), the characteristic function is (since $-\xi_r\xi_r = 1$, $-\overline{\xi}_r\overline{\xi}_r = 1$)

$$V(P', \overline{P}'; P, \overline{P}) = c(m + \overline{m})^{\frac{1}{2}}(mk^2 + \overline{m}\overline{k}^2 - 2mk\xi_r x_r - 2\overline{m}\overline{k}\overline{\xi}_r\overline{x}_r + \ldots)^{\frac{1}{2}},$$
(5.3.18)

the terms not written explicitly being finite for large k. Hence, expanding the radical, we have approximately

$$V(P', \overline{P}'; P, \overline{P}) = c(m + \overline{m})^{\frac{1}{2}}(mk^2 + \overline{m}\overline{k}^2)^{\frac{1}{2}}$$
$$- c\left(\frac{m + \overline{m}}{mk^2 + \overline{m}\overline{k}^2}\right)^{\frac{1}{2}}(mk\xi_r x_r + \overline{m}\overline{k}\overline{\xi}_r\overline{x}_r), \quad (5.3.19)$$

the omitted terms being small, of order $1/k$.

If we hold k, $\overline{k}$ constant, then $V = $ const. gives us the family of 7-waves from the source-event-pair (5.3.17). If we now regard this family as the family generated from the 7-wave Σ_0 which passes through the origin of H_8, the characteristic function $V(P, \overline{P})$ of this family differs from (5.3.19) only by a constant, viz. the action from $(P', \overline{P}')$ to Σ_0. Thus we have, approximately for large k,

$$V(P, \overline{P}) = -c\left(\frac{m + \overline{m}}{mk^2 + \overline{m}\overline{k}^2}\right)^{\frac{1}{2}}(mk\xi_r x_r + \overline{m}\overline{k}\overline{\xi}_r\overline{x}_r). \quad (5.3.20)$$

In the limit $k \to \infty$, in which P', $\overline{P}'$ have receded to infinity in space-time in directions opposed to ξ_r, $\overline{\xi}_r$ respectively, we have *accurately*

$$V(P, \overline{P}) = -c\left(\frac{m + \overline{m}}{m + \overline{m}\rho^2}\right)^{\frac{1}{2}}(m\xi_r x_r + \rho\overline{m}\overline{\xi}_r\overline{x}_r), \quad (5.3.21)$$

or

$$V(P, \overline{P}) = -\sigma_r x_r - \overline{\sigma}_r\overline{x}_r, \quad (5.3.22)$$

where

$$\sigma_r = c\left(\frac{m + \overline{m}}{m + \overline{m}\rho^2}\right)^{\frac{1}{2}}m\xi_r, \quad \overline{\sigma}_r = c\left(\frac{m + \overline{m}}{m + \overline{m}\rho^2}\right)^{\frac{1}{2}}\rho\overline{m}\overline{\xi}_r. \quad (5.3.23)$$

Here $\rho = \lim \overline{k}/k$. We thus obtain in the limit the set of plane 7-waves

$$\sigma_r x_r + \overline{\sigma}_r\overline{x}_r = \text{const.} \quad (5.3.24)$$

Since in the limit the straight world line $P'P$ has the direction of ξ_r and $\overline{P}'\overline{P}$ has that of $\overline{\xi}_r$, we recognize that ξ_r, $\overline{\xi}_r$ are the 4-velocities of the two particles, and so

$$\begin{aligned}
\xi_\rho &= \gamma v_\rho/c, & \xi_4 &= i\gamma, & \gamma &= (1 - v^2/c^2)^{-\frac{1}{2}}, \\
\overline{\xi}_\rho &= \overline{\gamma}\overline{v}_\rho/c, & \overline{\xi}_4 &= i\overline{\gamma}, & \overline{\gamma} &= (1 - \overline{v}^2/c^2)^{-\frac{1}{2}}.
\end{aligned} \quad (5.3.25)$$

Hence (5.3.23) gives the slowness 4-vectors in terms of the velocities as follows:

$$\left.\begin{array}{ll} \sigma_\rho = \left(\dfrac{m+\overline{m}}{m+\overline{m}\rho^2}\right)^{\frac{1}{2}} m\gamma v_\rho, & \sigma_4 = \left(\dfrac{m+\overline{m}}{m+\overline{m}\rho^2}\right)^{\frac{1}{2}} im\gamma c, \\[2ex] \overline{\sigma}_\rho = \left(\dfrac{m+\overline{m}}{m+\overline{m}\rho^2}\right)^{\frac{1}{2}} \rho\overline{m}\gamma v_\rho, & \overline{\sigma}_4 = \left(\dfrac{m+\overline{m}}{m+\overline{m}\rho^2}\right)^{\frac{1}{2}} \rho i\overline{m}\gamma c. \end{array}\right\} \quad (5.3.26)$$

The presence of ρ corresponds to the indeterminacy mentioned earlier. It is, however, now fixed as the limit of the ratio

$$P'O/\overline{P}'O = \overline{k}/k.$$

As a particular case (the most interesting one), let us make P' and $\overline{P}'$ go off into the remote past *in the same space-time direction*, remaining at a finite space-time distance from one another. As we are concerned only with the limit $k \to \infty$, this is the same as making $P', \overline{P}'$ coincide and go off to infinity together. We have then

$$\xi_r = \overline{\xi}_r, \quad \rho = 1, \tag{5.3.27}$$

and so, by (5.3.25), $v_\rho = \overline{v}_\rho$: *the particles have the same velocity.* Further, we have by (5.3.26)

$$\left.\begin{array}{ll} \sigma_\rho = m\gamma v_\rho, & \sigma_4 = im\gamma c, \\ \overline{\sigma}_\rho = \overline{m}\gamma v_\rho, & \overline{\sigma}_4 = i\overline{m}\gamma c, \end{array}\right\} \tag{5.3.28}$$

so that the slowness 4-vectors are in fact the usual momentum-energy 4-vectors of the two particles.

Let us sum up this result: *Given that the two particles had in their histories a common event (or, more generally, events at finite Min-kowskian distance from one another), situated far from the origin of space and in the remote past, then they travel through the finite portion of space during a finite range of time with a common velocity v_ρ, and these data determine a system of plane 7-waves with characteristic function*

$$V(P, \overline{P}) = -\sigma_r x_r - \overline{\sigma}_r \overline{x}_r, \tag{5.3.29}$$

where σ_r and $\overline{\sigma}_r$ are the momentum-energy 4-vectors of the particles, as shown in (5.3.28).

5.4. Interference of two particles passing through two holes

We do not get interference for two free particles in empty space-time because two event-pairs are connected by a single ray, and so there are not two wave systems to interfere with one another. Let us now consider the case where there is a screen with two holes in it. Two particles approach the screen from one side and pass through the holes. The space-time diagram is shown in fig. 5.2, where H_1, H_2 are the histories of the two holes, $(P', \bar{P}')$ the initial event-pair and $(P, \bar{P})$ the final event-pair.

The history of the pair of holes is given by six equations, and so this history is a 2-space (say S) in H_8. To construct the ray from $(P', \bar{P}')$ to $(P, \bar{P})$, we take a point in S, i.e. an event-pair $(Q, \bar{Q})$, with one event on H_1 and the other on H_2, and draw the rays in H_8

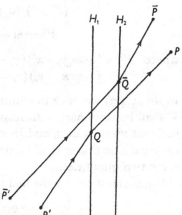

Fig. 5.2. Space-time diagram for the passage of two particles through two holes.

$$(P', \bar{P}') \text{ to } (Q, \bar{Q}),$$
$$(Q, \bar{Q}) \text{ to } (P, \bar{P}).$$

Each of these rays has an action of the form (5.3.7), and the action for the actual ray from $(P', \bar{P}')$ to $(P, \bar{P})$ through the holes is found by making the sum of these actions stationary with respect to variations of Q and $\bar{Q}$ on the world lines of the holes. Thus V can be calculated.

But there are actually *four* characteristic functions, corresponding to the different ways in which the particles can go through the holes. Fig. 5.2 shows only one case, viz. that in which the particle which has P', P in its history goes through H_1 and the other particle goes through H_2. The four possibilities are

(i) both particles through H_1,

(ii) both particles through H_2,

(iii) $P'P$ through H_1 and $\bar{P}'\bar{P}$ through H_2,

(iv) $P'P$ through H_2 and $\bar{P}'\bar{P}$ through H_1.

When we apply primitive quantization to the corresponding 7-waves, there will be interference.

The case of waves from a source-event-pair is rather complicated, and so we shall consider only the case of incident plane 7-waves generated from Σ_0 with equation

$$\sigma'_r x'_r + \bar{\sigma}'_r \bar{x}'_r = 0. \tag{5.4.1}$$

Denoting by y_r, $\bar{y}_r$ the coordinates of Q, $\bar{Q}$, we note that y_ρ, $\bar{y}_\rho$ are constants determined by the positions of the holes, which we take to be at rest in the observer's frame of reference. Then, by (5.3.7), the action along the broken line in H_8

$$(P', \bar{P}') \text{ to } (Q, \bar{Q}) \text{ to } (P, \bar{P})$$

is
$$V = c(m + \bar{m})^{\frac{1}{2}} (W' + W), \tag{5.4.2}$$

where
$$\begin{aligned} W' &= [-m(y_r - x_r')(y_r - x_r') - \bar{m}(\bar{y}_r - \bar{x}_r')(\bar{y}_r - \bar{x}_r')]^{\frac{1}{2}}, \\ W &= [-m(x_r - y_r)(x_r - y_r) - \bar{m}(\bar{x}_r - \bar{y}_r)(\bar{x}_r - \bar{y}_r)]^{\frac{1}{2}}. \end{aligned} \tag{5.4.3}$$

From (5.4.2) we have to eliminate x_r' and $\bar{x}_r'$ by the condition that V shall be stationary with respect to variation of P' and $\bar{P}'$ on the incident wave (5.4.1), and then we have to eliminate y_4 and $\bar{y}_4$ by the condition that V shall be stationary with respect to variations of these two quantities.

From variations on the incident wave we get

$$m(y_r - x_r') = \theta \sigma_r', \quad \bar{m}(\bar{y}_r - \bar{x}_r') = \theta \bar{\sigma}_r', \tag{5.4.4}$$

where θ is a positive undetermined factor. We substitute these values of σ_r', $\bar{\sigma}_r'$ in $\Omega = 0$, which as in (5.3.1) reads

$$\frac{1}{2m} \sigma_r' \sigma_r' + \frac{1}{2\bar{m}} \bar{\sigma}_r' \bar{\sigma}_r' + \tfrac{1}{2}(m + \bar{m}) c^2 = 0, \tag{5.4.5}$$

and so obtain, in the notation of (5.4.3),

$$-W'^2 + (m + \bar{m}) c^2 \theta^2 = 0. \tag{5.4.6}$$

Substitution for x_r', $\bar{x}_r'$ from (5.4.4) in (5.4.1) gives

$$\sigma_r'\left(y_r - \sigma_r' \frac{\theta}{m}\right) + \bar{\sigma}_r'\left(\bar{y}_r - \bar{\sigma}_r' \frac{\theta}{\bar{m}}\right) = 0, \tag{5.4.7}$$

or by (5.4.5)
$$\sigma_r' y_r + \bar{\sigma}_r' \bar{y}_r + \theta(m + \bar{m}) c^2 = 0. \tag{5.4.8}$$

With the value of θ so determined, (5.4.6) gives

$$W' = \theta c(m + \bar{m})^{\frac{1}{2}} = \frac{-\sigma_r' y_r - \bar{\sigma}_r' \bar{y}_r}{c(m + \bar{m})^{\frac{1}{2}}}, \tag{5.4.9}$$

and when we substitute this in (5.4.2) we get

$$V = -\sigma_r' y_r - \bar{\sigma}_r' \bar{y}_r + c(m + \bar{m})^{\frac{1}{2}} W. \tag{5.4.10}$$

We have now eliminated P', $\overline{P}'$. It remains to make V stationary with respect to y_4, $\overline{y}_4$, and the conditions for this are

$$\left.\begin{array}{l} -\sigma'_4 + c(m+\overline{m})^{\frac{1}{2}} \dfrac{m(x_4 - y_4)}{W} = 0, \\[2mm] -\overline{\sigma}'_4 + c(m+\overline{m})^{\frac{1}{2}} \dfrac{\overline{m}(\overline{x}_4 - \overline{y}_4)}{W} = 0. \end{array}\right\} \qquad (5.4.11)$$

Thus $\quad y_4 - x_4 = -\dfrac{W\sigma'_4}{cm(m+\overline{m})^{\frac{1}{2}}}, \quad \overline{y}_4 - \overline{x}_4 = -\dfrac{W\overline{\sigma}'_4}{c\overline{m}(m+\overline{m})^{\frac{1}{2}}}, \quad (5.4.12)$

and so by (5.4.3)

$$W^2 = -U^2 - \frac{W^2}{c^2(m+\overline{m})}\left(\frac{\sigma'^2_4}{m} + \frac{\overline{\sigma}'^2_4}{\overline{m}}\right), \qquad (5.4.13)$$

where $\quad U = [m(x_\rho - y_\rho)(x_\rho - y_\rho) + \overline{m}(\overline{x}_\rho - \overline{y}_\rho)(\overline{x}_\rho - \overline{y}_\rho)]^{\frac{1}{2}}. \quad (5.4.14)$

Hence $\quad W = U\left[-1 - \dfrac{1}{c^2(m+\overline{m})}\left(\dfrac{\sigma'^2_4}{m} + \dfrac{\overline{\sigma}'^2_4}{\overline{m}}\right)\right]^{-\frac{1}{2}}. \qquad (5.4.15)$

Now by (5.4.10) and (5.4.12)

$$V = -\sigma'_\rho y_\rho - \overline{\sigma}'_\rho \overline{y}_\rho - \sigma'_4\left(x_4 - \frac{W\sigma'_4}{cm(m+\overline{m})^{\frac{1}{2}}}\right)$$

$$\qquad\qquad -\overline{\sigma}'_4\left(\overline{x}_4 - \frac{W\overline{\sigma}'_4}{c\overline{m}(m+\overline{m})^{\frac{1}{2}}}\right) + c(m+\overline{m})^{\frac{1}{2}}W$$

$$= -\sigma'_\rho y_\rho - \overline{\sigma}'_\rho \overline{y}_\rho - \sigma'_4 x_4 - \overline{\sigma}'_4 \overline{x}_4$$

$$\qquad\qquad + Wc(m+\overline{m})^{\frac{1}{2}}\left[1 + \frac{1}{c^2(m+\overline{m})}\left(\frac{\sigma'^2_4}{m} + \frac{\overline{\sigma}'^2_4}{\overline{m}}\right)\right], \quad (5.4.16)$$

and so by (5.4.15) we finally have *the characteristic function for the refraction of plane 7-waves through two holes,*

$$V(P, \overline{P}) = -\sigma'_\rho y_\rho - \overline{\sigma}'_\rho \overline{y}_\rho - \sigma'_4 x_4 - \overline{\sigma}'_4 \overline{x}_4 - KU, \qquad (5.4.17)$$

where

$$\left.\begin{array}{l} K = \left[-(m+\overline{m})c^2 - \dfrac{\sigma'^2_4}{m} - \dfrac{\overline{\sigma}'^2_4}{\overline{m}}\right]^{\frac{1}{2}} = \left[\dfrac{\sigma'_\rho \sigma'_\rho}{m} + \dfrac{\overline{\sigma}'_\rho \overline{\sigma}'_\rho}{\overline{m}}\right]^{\frac{1}{2}}, \\[3mm] U = [m(x_\rho - y_\rho)(x_\rho - y_\rho) + \overline{m}(\overline{x}_\rho - \overline{y}_\rho)(\overline{x}_\rho - \overline{y}_\rho)]^{\frac{1}{2}}. \end{array}\right\} \quad (5.4.18)$$

V is a function of the eight variables x_r, $\overline{x}_r$, the coordinates of the final events, all else in (5.4.17) being known constants.

The interference pattern. Let the spatial coordinates of the holes be a_ρ for H_1 and b_ρ for H_2. In (5.4.17) y_ρ are the coordinates of the hole used by $P'P$ and $\overline{y}_\rho$ the coordinates of the hole used by $\overline{P}'\overline{P}$.

156   GEOMETRICAL MECHANICS

The four cases mentioned earlier then give the four characteristic functions as follows:

(i) both particles through H_1:
$$V_{aa} = -\sigma'_\rho a_\rho - \overline{\sigma}'_\rho a_\rho - \sigma'_4 x_4 - \overline{\sigma}'_4 \overline{x}_4 - K(mr_1^2 + \overline{m}\overline{r}_1^2)^{\frac{1}{2}};$$

(ii) both particles through H_2:
$$V_{bb} = -\sigma'_\rho b_\rho - \overline{\sigma}'_\rho b_\rho - \sigma'_4 x_4 - \overline{\sigma}'_4 \overline{x}_4 - K(mr_2^2 + \overline{m}\overline{r}_2^2)^{\frac{1}{2}};$$

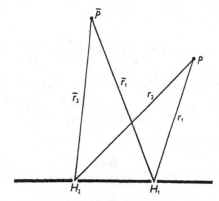

Fig. 5.3. Space diagram for interference of two particles passing through two holes.

(iii) $P'P$ through H_1 and $P'P$ through H_2:
$$V_{ab} = -\sigma'_\rho a_\rho - \overline{\sigma}'_\rho b_\rho - \sigma'_4 x_4 - \overline{\sigma}'_4 \overline{x}_4 - K(mr_1^2 + \overline{m}\overline{r}_2^2)^{\frac{1}{2}};$$

(iv) $P'P$ through H_2 and $P'P$ through H_1:
$$V_{ba} = -\sigma'_\rho b_\rho - \overline{\sigma}'_\rho a_\rho - \sigma'_4 x_4 - \overline{\sigma}'_4 \overline{x}_4 - K(mr_2^2 + \overline{m}\overline{r}_1^2)^{\frac{1}{2}}.$$

In these formulae
$$\begin{cases} r_1 = [(x_\rho - a_\rho)(x_\rho - a_\rho)]^{\frac{1}{2}}, \\ r_2 = [(x_\rho - b_\rho)(x_\rho - b_\rho)]^{\frac{1}{2}}, \\ \overline{r}_1 = [(\overline{x}_\rho - a_\rho)(\overline{x}_\rho - a_\rho)]^{\frac{1}{2}}, \\ \overline{r}_2 = [(\overline{x}_\rho - b_\rho)(\overline{x}_\rho - b_\rho)]^{\frac{1}{2}}, \end{cases} \tag{5.4.19}$$

these being the spatial distances of the final events P, $\overline{P}$ from the two holes, as shown in fig. 5.3.

We now introduce primitive quantization, adjacent γ-waves of equal phase satisfying the condition that the action between them is h (Planck's constant). A 'bright point' in H_8 occurs where all four waves are in phase, i.e. where the differences between the four quantities

$$V_{aa}, \quad V_{bb}, \quad V_{ab}, \quad V_{ba}$$

are integer multiples of h. Accordingly we have for bright points the three equations

$$
\left.
\begin{aligned}
V_{aa} - V_{bb} &= \sigma'_\rho(b_\rho - a_\rho) + \overline{\sigma}'_\rho(b_\rho - a_\rho) + K(mr_2^2 + \overline{m}\overline{r}_2^2)^{\frac{1}{2}} \\
&\quad - K(mr_1^2 + \overline{m}\overline{r}_1^2)^{\frac{1}{2}} = nh, \\
V_{aa} - V_{ab} &= \overline{\sigma}'_\rho(b_\rho - a_\rho) + K(mr_1^2 + \overline{m}\overline{r}_2^2)^{\frac{1}{2}} \\
&\quad - K(mr_1^2 + \overline{m}\overline{r}_1^2)^{\frac{1}{2}} = n'h, \\
V_{aa} - V_{ba} &= \sigma'_\rho(b_\rho - a_\rho) + K(mr_2^2 + \overline{m}\overline{r}_1^2)^{\frac{1}{2}} \\
&\quad - K(mr_1^2 + \overline{m}\overline{r}_1^2)^{\frac{1}{2}} = n''h,
\end{aligned}
\right\} \quad (5.4.20)
$$

where $n,\ n',\ n'' = 0,\ \pm 1,\ \pm 2,\ \ldots$.

Let us for simplicity suppose that the incident waves satisfy

$$\sigma'_\rho(b_\rho - a_\rho) = 0, \quad \overline{\sigma}'_\rho(b_\rho - a_\rho) = 0. \qquad (5.4.21)$$

We shall interpret these conditions later in a special case. Then (5.4.20) simplify to

$$
\left.
\begin{aligned}
(mr_2^2 + \overline{m}\overline{r}_2^2)^{\frac{1}{2}} - (mr_1^2 + \overline{m}\overline{r}_1^2)^{\frac{1}{2}} &= nh/K, \\
(mr_1^2 + \overline{m}\overline{r}_2^2)^{\frac{1}{2}} - (mr_1^2 + \overline{m}\overline{r}_1^2)^{\frac{1}{2}} &= n'h/K, \\
(mr_2^2 + \overline{m}\overline{r}_1^2)^{\frac{1}{2}} - (mr_1^2 + \overline{m}\overline{r}_1^2)^{\frac{1}{2}} &= n''h/K.
\end{aligned}
\right\} \quad (5.4.22)
$$

We note that x_4 and $\overline{x}_4$ do not appear in (5.4.20) or (5.4.22). That means that the interference pattern is fixed, as indeed we are to expect for plane incident waves. Thus a bright point of H_8 becomes a pair of points in the space of the observer. For assigned integer values of $n,\ n',\ n''$, (5.4.22) are three equations for the four distances $r_1,\ r_2,\ \overline{r}_1,\ \overline{r}_2$.

Let us now approximate, taking h small. If $h = 0$, then (5.4.22) imply

$$r_1 = r_2, \quad \overline{r}_1 = \overline{r}_2, \qquad (5.4.23)$$

and so, if we define $r_0,\ \overline{r}_0,\ \epsilon,\ \overline{\epsilon}$ by

$$
\left.
\begin{aligned}
r_2 &= r_0 + \epsilon, \quad r_1 = r_0 - \epsilon, \\
\overline{r}_2 &= \overline{r}_0 + \overline{\epsilon}, \quad \overline{r}_1 = \overline{r}_0 - \overline{\epsilon},
\end{aligned}
\right\} \quad (5.4.24)
$$

then, for small h, ϵ and $\overline{\epsilon}$ will be small. Accordingly (5.4.22) gives approximately

$$
\left.
\begin{aligned}
mr_0\epsilon + \overline{m}\overline{r}_0\overline{\epsilon} = \frac{1}{2}\frac{nhL}{K}, \quad \overline{m}\overline{r}_0\overline{\epsilon} = \frac{1}{2}\frac{n'hL}{K}, \quad mr_0\epsilon = \frac{1}{2}\frac{n''hL}{K}, \\
L = (mr_0^2 + \overline{m}\overline{r}_0^2)^{\frac{1}{2}}.
\end{aligned}
\right\} \quad (5.4.25)
$$

Hence, in the approximation, the quantum integers are not independent, for (5.4.25) implies

$$n = n' + n'', \tag{5.4.26}$$

and so, instead of *three* conditions we now have only *two*:

$$mr_0\epsilon = \frac{1}{2}\frac{n''hL}{K}, \quad \overline{m}\bar{r}_0\bar{\epsilon} = \frac{1}{2}\frac{n'hL}{K}. \tag{5.4.27}$$

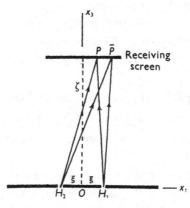

Fig. 5.4. Reception of two particles on a screen.

We recall that K is, as in (5.4.18), a constant defined by the incident 7-waves.

The interference pattern has symmetry of rotation about the line joining the holes, and so we can study the pattern by confining ourselves to a plane containing the holes. We take the holes at $(\xi, 0, 0)$ and $(-\xi, 0, 0)$ and study the plane $x_2 = 0$ (fig. 5.4). Since $\epsilon, \bar{\epsilon}$ are small, we confine our attention to the neighbourhood of the x_3-axis.

Now approximately

$$\epsilon = \xi x_1/r_0, \quad \bar{\epsilon} = \xi\bar{x}_1/\bar{r}_0, \tag{5.4.28}$$

and so the conditions (5.4.27) read

$$x_1 = \frac{1}{2}\frac{n''hL}{\xi mK}, \quad \bar{x}_1 = \frac{1}{2}\frac{n'hL}{\xi \overline{m}K}, \quad L = (mr_0^2 + \overline{m}\bar{r}_0^2)^{\frac{1}{2}}. \tag{5.4.29}$$

If we now set a receiving screen at $x_3 = \zeta$ and make ζ/ξ large, we have approximately

$$r_0 = \bar{r}_0 = \zeta, \quad L = \zeta(m + \overline{m})^{\frac{1}{2}}, \tag{5.4.30}$$

and so (5.4.29) gives

$$x_1 = \frac{\zeta}{2\xi} \frac{n''h(m+\overline{m})^{\frac{1}{2}}}{mK}, \quad \overline{x}_1 = \frac{\zeta}{2\xi} \frac{n'h(m+\overline{m})^{\frac{1}{2}}}{\overline{m}K}. \quad (5.4.31)$$

We see that the patterns are independent: there are two sets of bright points on the linear screen (parallel lines on a plane screen in three dimensions), one set for each particle. However, each distribution is influenced by the other, in the sense that x_1 depends on $\overline{m}$ and $\overline{x}_1$ depends on m.

Finally let us take the situation described at (5.3.29), in which the two particles had a common event in their history and arrive at the screen with the same velocity v_ρ. Then, by (5.3.28) and (5.4.18), we have

$$K = \left(\frac{\sigma'_\rho \sigma'_\rho}{m} + \frac{\overline{\sigma}'_\rho \overline{\sigma}'_\rho}{\overline{m}} \right)^{\frac{1}{2}} = (m+\overline{m})^{\frac{1}{2}} \gamma v, \quad (5.4.32)$$

and consequently (5.4.31) become

$$x_1 = \frac{\zeta}{2\xi} \frac{n''h}{m\gamma v}, \quad \overline{x}_1 = \frac{\zeta}{2\xi} \frac{n'h}{\overline{m}\gamma v}. \quad (5.4.33)$$

Each of these is precisely the formula (4.6.20) for a single particle refracted through two holes, and so *the effects due to the two particles have been completely separated*. Since σ'_ρ and $\overline{\sigma}'_\rho$ now have the spatial direction of v_ρ, the special simplifying conditions (5.4.21) mean that the common velocity of the two particles is perpendicular to the line joining the holes, as in the case of normal incidence.

Let us sum up: when plane 7-waves corresponding to two free particles fall on a screen containing two holes, under the simplifying condition (5.4.21), the interference pattern on a receiving screen is given by (5.4.31) under appropriate approximations. There are two sets of bright lines (more properly, lines of high probability), one set for each particle, the spacing of the lines of one set being influenced by the presence of the other particle in general. But if the two particles had a common event far off in the remote past, and so arrive with a common velocity, this mutual influence disappears, and we get the individual patterns as in (5.4.33), one for each particle as if it were alone.

5.5. The optical method and dynamical method

The purpose of this section is to describe, more fully than has been done on pp. 15, 16, the connexion between the optical and dynamical methods of Hamilton, as applied to relativistic mechanics. (Note that we are concerned solely with *mechanics*; relativistic *optics* is not touched on in this book.) The dynamical method is actually only a particular application of the optical method, and so the clearest way to see the connexion between them is to start from the optical method and trace the specialization by which the dynamical method is derived.

Let us take the general point of view as developed in §5.1, and start from the slowness equation (5.1.1):

$$\Omega(\sigma, x) = 0. \tag{5.5.1}$$

The rays satisfy (5.1.6):

$$\frac{dx^A}{dw} = \frac{\partial \Omega}{\partial \sigma_A}, \quad \frac{d\sigma_A}{dw} = -\frac{\partial \Omega}{\partial x^A}, \tag{5.5.2}$$

where $A = 1, 2, ..., N$. These equations (5.5.1), (5.5.2) may be said to summarize the optical method.

The essence of the dynamical method lies in solving (5.5.1) for one of the σ's, say σ_N; let us write the solution as

$$\sigma_N = K(\sigma_1, ... \sigma_{N-1}; x^1, ... x^N). \tag{5.5.3}$$

Now (5.5.2) always gives the rays, no matter in which of the infinitely many equivalent forms the slowness equation (5.5.1) is written, the only difference being a change in the parameter w. Let us then write (5.5.1) in the equivalent form

$$\Omega^*(\sigma, x) = kK(\sigma_1, ... \sigma_{N-1}; x^1, ... x^N) - k\sigma_N = 0, \tag{5.5.4}$$

the constant factor k being introduced merely for later notational convenience. The rays satisfy (5.5.2) with Ω changed to Ω^* and the parameter w changed to some new parameter w^*. Greek capitals taking the range $1, 2, ... N-1$, the equations of the rays now read

$$\frac{dx^\Gamma}{dw^*} = k\frac{\partial K}{\partial \sigma_\Gamma}, \quad \frac{d\sigma_\Gamma}{dw^*} = -k\frac{\partial K}{\partial x^\Gamma}, \tag{5.5.5}$$

$$\frac{dx^N}{dw^*} = -k, \quad \frac{d\sigma_N}{dw^*} = -k\frac{\partial K}{\partial x^N}. \tag{5.5.6}$$

Thus, with omission of an immaterial additive constant, the co-ordinate x^N is connected with the parameter w^* by

$$x^N = -kw^*. \qquad (5.5.7)$$

We now define a (Hamiltonian) function H by

$$H(\sigma_1, \ldots \sigma_{N-1}; x^1, \ldots x^{N-1}; w^*) = kK(\sigma_1, \ldots \sigma_{N-1}; x^1, \ldots x^N),$$
$$(5.5.8)$$

and the equations (5.5.5) read

$$\frac{dx^\Gamma}{dw^*} = \frac{\partial H}{\partial \sigma_\Gamma}, \quad \frac{d\sigma_\Gamma}{dw^*} = -\frac{\partial H}{\partial x^\Gamma}. \qquad (5.5.9)$$

The dynamical method is summarized by the prescription of the Hamiltonian function H and the equations (5.5.9) for the rays. To get complete formal agreement between (5.5.8), (5.5.9) and the usual presentation of Hamiltonian dynamics, we may put $\sigma_\Gamma = p_\Gamma$ (components of momentum) and $w^* = t$ (the time), although this last would be confusing in applications to problems involving several particles, since each has its own time.

The equations (5.5.1), (5.5.2) on the one hand, and the equations (5.5.8), (5.5.9) on the other, represent two equivalent ways of looking at the same mathematical situation. The former constitute the optical method and the latter the dynamical method. The difference is rather trivial mathematically: in the optical method we leave the slowness equation (5.5.1) unsolved, whereas in the dynamical method we solve it as in (5.5.3).

We might say that the optical method consists in using the slowness equation in the general form

$$\Omega(\sigma_1, \ldots \sigma_N; x^1, \ldots x^N) = 0, \qquad (5.5.10)$$

while the dynamical method consists in using it in the special form

$$\sigma_N - \omega(\sigma_1, \ldots \sigma_{N-1}; x^1, \ldots x^N) = 0. \qquad (5.5.11)$$

Viewed in this way, the optical method includes the dynamical method as a special case. We may prefer on occasion to use (5.5.11), just as in geometry we may prefer to describe a surface by $z - f(x, y) = 0$ rather than by $F(x, y, z) = 0$; but the latter includes the former.

To avoid falsifying history, I should remark that Hamilton did not give to his optical method the full power of generality, for he imposed on Ω in (5.5.10) the condition of homogeneity mentioned on p. 10. Thus what is called 'Hamilton's optical method' in this book is a slight generalization of the method he actually employed. Slight as it is, it seems important for the following reasons: first, it permits us to avoid normalizing Ω by the condition of homogeneity, and, secondly, it enables us to view the dynamical method as a particular case of the optical method. The form (5.5.11), essential in the dynamical method, is not in general consistent with the homogeneity condition imposed by Hamilton.

Let us now illustrate the connexion between the two methods in the case of a charged particle in an electromagnetic field, as discussed in §3.4. We have the slowness equation (3.4.11), which we shall write

$$\Omega(\sigma, x) = \tfrac{1}{2}(\sigma_r - A_r)(\sigma_r - A_r) + \tfrac{1}{2}m^2c^2 = 0. \qquad (5.5.12)$$

This is the equation (5.5.1) for this case. As in (5.5.2), we have for the rays

$$\frac{\mathrm{d}x_r}{\mathrm{d}w} = \frac{\partial\Omega}{\partial\sigma_r} = \sigma_r - A_r, \quad \frac{\mathrm{d}\sigma_r}{\mathrm{d}w} = -\frac{\partial\Omega}{\partial x_r} = (\sigma_n - A_n)\frac{\partial A_n}{\partial x_r}. \qquad (5.5.13)$$

These equations (5.5.12), (5.5.13) belong to the optical method. To use the dyanmical method, we solve (5.5.12) for σ_4, obtaining as in (5.5.3)

$$\sigma_4 = K(\sigma_1, \sigma_2, \sigma_3; x_1, x_2, x_3, x_4)$$
$$= A_4 \pm (2\sigma_\rho A_\rho - \sigma_\rho \sigma_\rho - A_\rho A_\rho - m^2c^2)^{\frac{1}{2}}. \qquad (5.5.14)$$

Let us choose $k = -ic$, so that (5.5.7) gives $ict = icw^*$, $w^* = t$. Then the Hamiltonian H of (5.5.8) is (with p_ρ written for σ_ρ)

$$H(p_1, p_2, p_3; x_1, x_2, x_3; t) = -ic\sigma_4$$
$$= -icA_4 \pm c(m^2c^2 + \sigma_\rho \sigma_\rho - 2\sigma_\rho A_\rho + A_\rho A_\rho)^{\frac{1}{2}}, \qquad (5.5.15)$$

and the equations of motion (5.5.9) read

$$\frac{\mathrm{d}x_\rho}{\mathrm{d}t} = \frac{\partial H}{\partial p_\rho}, \quad \frac{\mathrm{d}p_\rho}{\mathrm{d}t} = -\frac{\partial H}{\partial x_\rho}. \qquad (5.5.16)$$

The obligation to solve (5.5.12) for σ_4, imposed by the dynamical method, forces us to use the complicated expression (5.5.15) for H instead of dealing with (5.5.12) in its original simple form.

With regard to Lorentz invariance, essential in relativistic discussions, the optical method assures it by the choice of an invariant medium-function $f(x, \alpha)$ or an invariant function $\Omega(\sigma, x)$. Note the obvious invariance of (5.5.12). In the dynamical method the requirement is a little more subtle; H must transform like the fourth component of a 4-vector. We note that in (5.5.15) we have $H = -ic\sigma_4$, fulfilling this requirement.

REFERENCES

BACHER, R. F. and GOUDSMIT, S. (1932). *Atomic Energy States*. New York and London: McGraw-Hill Book Company.

BAKER, B. B. and COPSON, E. T. (1950). *The Mathematical Theory of Huygens' Principle*. Oxford: Clarendon Press.

CARATHÉODORY, C. (1935). *Variationsrechnung*. Berlin.

DE BROGLIE, L. and BRILLOUIN, L. (1928). *Selected Papers on Wave Mechanics*. London and Glasgow: Blackie.

DIRAC, P. A. M. (1930). *The Principles of Quantum Mechanics*. Oxford: Clarendon Press.

DIRAC, P. A. M. (1951). *Proc. Roy. Soc.* A, **209**, 291.

HADAMARD, J. (1910). *Leçons sur le Calcul des Variations*, tome 1. Paris: Hermann.

HAMILTON, Sir W. R. (1931). *Mathematical Papers*, vol. 1. Cambridge University Press.

HAMILTON, Sir W. R. (1941). *Mathematical Papers*, vol. 2. Cambridge University Press.

RAYLEIGH, LORD (1877). *Proc. London Math. Soc.* **9**, 21; cf. *Theory of Sound*, **1**, 475 (Dover Publications, New York, 1945).

SOMMERFELD, A. (1934). *Atomic Structure and Spectral Lines*, vol. 1. London: Methuen.

SYNGE, J. L. (1937). *Geometrical Optics*. Cambridge University Press.

SYNGE, J. L. (1951). *Hamilton's Method in Geometrical Optics*. University of Maryland. Lecture Series, no. 9.

SYNGE, J. L. (1952). *Proc. Roy. Soc.* A, **211**, 303.

SYNGE, J. L. (1953). *Physical Review*, **89**, 467.

WATSON, G. N. (1944). *Theory of Bessel Functions*. Cambridge University Press.

INDEX